AF678823

Reliability and Maintainability Management

by

Balbir S. Dhillon
Department of Mechanical Engineering
University of Ottawa

and

Hans Reiche
Reliability and Maintainability Consultant, Ottawa

VNR VAN NOSTRAND REINHOLD COMPANY
New York

Copyright © 1985 by Van Nostrand Reinhold Company Inc.

Library of Congress Catalog Card Number: 84-25742
ISBN: 0-442-27637-0

All rights reserved. No part of this work covered by the copyright hereon may be reproduced or used in any form or by any means—graphic, electronic, or mechanical, including photocopying, recording, taping, or information storage and retrieval systems—without permission of the publisher.

Manufactured in the United States of America

Published by Van Nostrand Reinhold Company Inc.
135 West 50th Street
New York, New York 10020

Van Nostrand Reinhold Company Limited
Molly Millars Lane
Workingham, Berkshire RG11 2PY, England

Van Nostrand Reinhold
480 Latrobe Street
Melbourne, Victoria 3000, Australia

Macmillan of Canada
Division of Gage Publishing Limited
164 Commander Boulevard
Agincourt, Ontario M1S 3C7, Canada

15 14 13 12 11 10 9 8 7 6 5 4 3 2 1

Library of Congress Cataloging in Publication Data

Dhillon, B. S.
Reliability and maintainability management.

Includes index.
1. Reliability (Engineering) 2. Maintainability (Engineering) I. Reiche, Hans. II. Title.
TS173.D49 1985 658.5′62 84-25742
ISBN 0-442-27637-0

This book is affectionately dedicated to the memory of our late beloved fathers, Karnail S. Dhillon and Prof. Dr. Fritz Reiche.

Preface

The aim of this book is to give engineers and managers some guidance in the areas of reliability and maintainability management. In the past, engineers and designers had little guidance from management in the areas of reliability and maintainability. Management has had limited practical help from anyone to assist in its decision-making process. The bottom line remains cost, and few managers have tied this factor to reliability or maintainability. Lack of suitable guidance for management is recognized as having a major influence on the reliability or maintainability of any system or equipment. Lessons learned over the last decade indicate that more management awareness must take place, and that involvement during the life cycle is necessary to improve system readiness and reduce life cycle cost.

The life cycle of a system or a piece of equipment sets the stage for determining the responsibilities and tasks that various levels and lines of management should follow. The book discusses these responsibilities and indicates the kind of reliability and maintainability tools, and the kind of technical documentation with which each level must be competent. The book eludes to the controls that must be installed, and the need for management information, which must be made available for effective decision-making and guidance purposes. The controls mentioned are in the form of specific actions and decisions by management. The decisions which are to be made for proposals, reviews, changes, funding, manpower, and timing are discussed. Life cycle costing and contract warranties are discussed in the context of effective reliability and maintainability management. Many other major subjects with which management must become familiar, are also outlined in the text. The contents of each chapter are described briefly in Chapter 1.

The mathematical contents of the book are presented with the aid of many solved examples. References to the relevant literature for almost all chapters are provided for readers. This will provide a useful service to readers when additional reading is necessary. Generally, the inclination of the book is on the structure of concepts rather than on minute details.

The book can easily be understood by anyone with a certain knowledge in engineering or related fields. The book will be useful to engineering managers and administrators, reliability and maintainability engineers, maintenance engineers, systems engineers, and others. In addition, it can be used to teach undergraduate, graduate, and professional courses related to, or specifically on, reliability and maintainability management. The information is written in a format that allows students to learn and understand the philosophy of reliability and maintainability management better and, at the same time, tests their comprehension through exercises listed at the end of each chapter.

We wish to thank many professionals who, through discussion and writing, have influenced our thinking. In particular, the assistance of Mr. A.P. Harris and Mr. R. Bezant is highly appreciated, for the development and writing of chapter 10 and chapter 12, respectively. We would also like to thank Mr. R. Buch and Mrs. Rosy Dhillon for editing and word processing the manuscript.

BALBIR S. DHILLON
HANS REICHE
Ottawa, Ontario

Contents

CHAPTER 1

Introduction

1.1. INTRODUCTION

There can be no doubt that only a few reliability and maintainability techniques have found their way into the upper management decision-making process. This is not surprising because management is concerned with profit and cost, and the relationship of these factors to reliability or maintainability is often not fully understood. Not only is there limited understanding of these concepts, but there is almost no clear and practical direction or guide available which suggests the kind of tasks that management can use to assist them.

Any project that deals with reliability and maintainability requires that certain steps be taken to achieve the required results. The speed with which technological advances are being made are almost exponential in nature, and these advances require careful management consideration. These advances bring with them new management problems which have not been encountered before and difficulties are being experienced in solving them, especially when the cost factor rises continuously at the end of the line. The advice given to management by other levels concerning reliability and maintainability is often not translatable enough for management to take action, and the two levels become more and more separated in their understanding of the problem. Management is sometimes impressed with the reliability engineering jargon to the extent that a simple statement of a high mean time between failures (MTBF) value means a major cost reduction for them. Little consideration is given to the fact that there are many other factors that must be taken into account when calculating cost reductions. The question that is raised here is, where are the guides for any tasks with which management should consult, for its level of operation?

1.2. MANAGEMENT TASKS

One would think that there are enough books written that explain the reasons why cost is influenced by reliability and maintainability techniques, and moreover, that management is making good use of such books for its decisions. While it is true that most of these books deal with the question of how reliability and maintainability affects cost, there is little guidance as to what exactly management must do. This book attempts to assist various levels of management to deal with reliability and maintainability problems, to explain some of the basic technologies in simple terms, and to improve the understanding of these factors, which have a direct influence on management decisions.

For the purpose of introducing management tasks, the basic mathematical forms are explained together with a general introduction of the concept of reliability and maintainability. The aspect of management is addressed and its related manpower is introduced. The suggested activities for various levels and lines of management are listed and cover both hardware and software. The various tools that a manager requires to carry out his tasks are also mentioned. One of the key issues is the manpower requirement, and a number of tasks are identified which are directly related to this high, life cycle cost item. Making data available for any decision is important and an outline is given on how this data can be obtained. Life cycle costing and warranties are powerful tools which are available to management, and the types of calculations which can be made are explained. Both techniques play a major role in cost control, assessment, and reduction. Hardware-software changes are always made during the life cycle. Reviews and engineering change proposals require certain management tasks and these are outlined.

1.3. SCOPE

The scope of reliability engineering is extremely wide, encompassing many areas of engineering technology. Reliability engineering helps ensure the success of space missions, maintain the national security, deliver a steady supply of electric power, provide reliable transportation, etc. There has been a considerable growth of knowledge in several areas of reliability engineering and its applications. So far, over one hundred books on the various aspects of reliability engineering have been published. However, to our best knowledge, there is no book that covers the subject of reliability and maintainability management within the framework of a single volume.

The topic of reliability and maintainability management is equally as important as the technical aspect of reliability and maintainability engineering. Today, more emphasis is being given to the managerial aspect of reliability in both the industrial and academic world. For example, IEEE Transactions on Reliability (a journal) devoted one issue to the Management of Reliability Programs.

Up to the present, engineers and designers of equipment and systems have had little guidance from management in areas of reliability and maintainability. Management has had little practical help for decision-making in regard to these subjects. For management, the bottom line is cost and few managers have tied this factor to reliability or maintainability. Lack of suitable management decisions and lack of guidance to others is recognized to have an influence on reliability and maintainability, and with it, cost. There are certain responsibilities and tasks that managers should be aware of and exercise to influence the bottom line. Many lessons have been learned over the last two decades which indicate that management must get involved during the life cycle of an equipment and a system in order to improve readiness and reduce life cycle cost.

This book covers the various management tasks that must be carried out for any successful reliability and maintainability project. Emphasis is on what tasks to do and how to do them, rather than why to do them. Complex mathematical models are avoided, and only information that is necessary for evaluation, assessment, or decision-making is presented. Because this is not a reliability and maintainability standard textbook, only those parameters which directly or indirectly influence management decisions are addressed. Moreover, the information contained in this book can be used by all management levels and by all engineers to acquaint them with the management interface and the required support for a project.

The topics covered in this book are of current interest, and the market for the book will be very large (considering the number of people employed in product development and procurement fields). The book will be useful to all types of practicing engineers, engineering managers and administrators, reliability and maintainability engineers, maintenance engineers, systems engineers, and undergraduate and graduate students.

The information is written in a format that allows students to learn and better understand the philosophy of reliability and maintainability management, and at the same time, test their comprehension through listed exercises.

1.4 DESCRIPTION OF TEXT

The activities of reliability and maintainability management are covered in twelve chapters, excluding the Introduction Chapter.

Chapter 2 is concerned with basic reliability and maintainability mathematical concepts. These concepts are considered useful in understanding the subsequent chapters of this book. Chapter 2 explains the basic rules of differentiation and integration, the quadratic equation, the fundamental axioms of probability, the conditional probability, three failure density functions, etc.

The basic concepts of reliability and maintainability are covered in Chapter 3, and histories of reliability and maintainability are briefly discussed. Selective definitions and sources of current reliability and maintainability information are presented, along with well-known bathtub hazard rate curve. Reliability networks such as series, parallel, the standby system, and *r*-out-of-*m* units are presented. Formulas to obtain component reliability, mean time to failure, and hazard rate are briefly discussed, and the maintainability function is defined. System steady state availability and unavailability formulas are presented. The hazard plotting method to estimate values of parameters of the Weibull distribution is briefly described.

Chapter 4 discusses various aspects of general management including the history of management, methods of organization, organization planning, span of supervision, and organization diagrams. Functions or responsibilities of reliability engineering, maintenance engineering, and quality control departments are outlined. The method of group brainstorming is discussed briefly, and some management work-related ratios and indices are presented.

The subject of engineering manpower is covered in Chapter 5. The Chapter presents the qualities, qualifications, attributes, characteristics, functions, etc., of selective professionals such as engineering managers, project engineers, reliability engineers, human factors engineers, and quality control managers.

Chapter 6 covers the life cycle management of a system. The life cycle consists of four different phases and each phase requires certain management action. The tasks that must be performed during the life cycle are explained, as well as the constraints that must be considered. Different management levels are listed whose members must perform certain tasks during the life cycle.

Chapter 7 covers the activities of reliability and maintainability management. This chapter is based on the previous chapter and relates the activities of the life cycle. Detailed suggestions regarding these various activities are made that allow management to assess the impact that such activities have on any reliability and maintainability project, and thus determine how much effort must be spent on these activities.

Chapter 8 discusses the management of reliability and maintainability documents. A large number of documents that arise from the various life cycle management phases and associated activities must be dealt with by management. The cost of documentation is a major life cycle cost item and certainly is of interest to any management. The types of documents that must be dealt with, and the application these documents have during the life cycle are explained. Configuration management is covered by the controls, status accounts, audits, and review activities.

Chapter 9 is concerned with the management tools for reliability managers and addresses four important areas: procedures for reliability assurance, configuration management, value engineering, and network analysis technique.

Chapter 10 reviews the reliability and maintenance manpower requirements as a major life cycle cost contributor. The various factors that contribute to the manpower requirements are shown. These factors are derived from reliability and maintainability information collected during the life cycle phases.

Chapter 11 discusses the reliability and maintainability management data needs; what kind of data can be made available to management; who needs the data; when the data is required; and other factors. Management's function in the dissemination and use of data for decision-making purposes is outlined, and the importance of data and the fallacies of data overkill are mentioned.

Chapter 12 outlines reliability and maintainability review and design change processes. Many different types of reviews are listed and explained, and the relationship to the life cycle and the timing of reviews or design changes are indicated.

The topics of life cycle costing and warranties are covered in Chapter 13. More specifically, life cycle cost, use, and planning are described, and the disciplines and activities associated with life cycle cost analysis are discussed. Also included are five life cycle cost models and ten mathematical models to estimate cost and return on investment. The topic of warranties is briefly covered.

The main objective of this book is to provide a reference document to management and others, which will foster better reliability and maintainability decision-making. Presently, there is no such book on the market to aid management in reliability and maintainability decision-making. Therefore, this book is being written for the first time to fill this vital gap.

CHAPTER 2

Reliability and Maintainability Mathematics

2.1. INTRODUCTION

In reliability and maintainability engineering mathematics is an indispensable tool. Hence, each person associated with reliability and maintainability analytical work must possess some knowledge of essential mathematics. The main purpose of this chapter is to familiarize readers with certain mathematical concepts useful in understanding subsequent chapters of the book. Some of the mathematical concepts are presented below.

2.2. DIFFERENTIATION, INTEGRATION, AND THE QUADRATIC EQUATION

All these items are described below, separately.

2.2.1. Differentiation

The process of finding a derivative is known as differentiation. If $M = f(t)$, the derivative of M with respect to variable t is defined as follows:

$$\frac{dM}{dt} = \lim_{\Delta t \to 0} \frac{f(t + \Delta t) - f(t)}{\Delta t} \tag{2.1}$$

There are several ways of denoting the derivative of M at t. Two of the examples are as follows:

i) M' (M prime).

ii) $\dfrac{d[f(t)]}{dt}$ (dee $f(t)$, dee t).

Differentiation Rules

Rules associated with differentiation are as follows:

RULE I

If

$$M = f(t) = t^k,$$

then

$$\frac{dM}{dt} = kt^{k-1}$$

where

k denotes any real number.

EXAMPLE 2.1

Prove that

$$\frac{dM}{dt} = kt^{k-1} \tag{2.2}$$

Since $M = t^k$ with the aid of the definition of the derivative, we get

$$\frac{dM}{dt} = \lim_{\Delta t \to 0} \frac{(t + \Delta t)^k - t^k}{\Delta t} \tag{2.3}$$

The expansion of the term $(t + \Delta t)^k$ can be written as

$$(t + \Delta t)^k = t^k + kt^{k-1} \cdot \Delta t + (\quad)t^{k-2}(\Delta t)^2 + \cdots + (\Delta t)^k \tag{2.4}$$

Substituting equation (2.4) into equation (2.3), results in

$$\frac{dM}{dt} = \lim_{\Delta t \to 0} \frac{t^k + kt^{k-1} \cdot \Delta t + (\quad)t^{k-2}(\Delta t)^2 + \cdots + (\Delta t)^k - t^k}{\Delta t} \tag{2.5}$$

From the above equation we get

$$\frac{dM}{dt} = \lim_{\Delta t \to 0} [kt^{k-1} + (\quad)t^{k-2} \cdot \Delta t + \cdots + (\Delta t)^{k-1}] \tag{2.6}$$

As $\Delta t \to 0$, the right hand term of equation (2.6) reduces to

$$\frac{dM}{dt} = kt^{k-1} \tag{2.7}$$

It proves that equation (2.7) is same as equation (2.2).

RULE II

If $M = f(t) = A$, then the derivative of M is equal to zero, or

$$\frac{dM}{dt} = 0$$

where

A is a constant.

EXAMPLE 2.2

Prove that

$$\frac{dM}{dt} = 0 \tag{2.8}$$

where

$$M = A$$

Since $M = A$, applying the definition of the derivative we get

$$\begin{aligned}\frac{dM}{dt} &= \lim_{\Delta t \to 0} \frac{f(t + \Delta t) - f(t)}{\Delta t} \\ &= \lim_{\Delta t \to 0} \frac{(A - A)}{\Delta t} \\ &= \lim_{\Delta t \to 0} \frac{0}{\Delta t} = \lim_{\Delta t \to 0} 0 = 0\end{aligned} \tag{2.9}$$

Thus,

$$\frac{dM}{dt} = 0, \tag{2.10}$$

which proves that equation (2.10) is same as equation (2.8).

RULE III

If $M = A \cdot f(t)$, then the derivative of M is

$$\frac{dM}{dt} = A \cdot \frac{d[f(t)]}{dt} \tag{2.11}$$

where

A is a constant.

EXAMPLE 2.3

If $M = 6t^4$, find $\dfrac{dM}{dt}$.

With the aid of equation (2.11), we get

$$\frac{dM}{dt} = 6 \cdot \frac{d[t^4]}{dt} = (6)(4)t^{4-1} = 24t^3$$

where

$$A = 6 \qquad \text{and} \qquad f(t) = t^4$$

RULE IV

If $M = f(t) + N(t)$, then

$$\frac{dM}{dt} = \frac{d[f(t)]}{dt} + \frac{d[N(t)]}{dt} \tag{2.12}$$

where

$f(t)$ and $N(t)$ are differentiable functions.

EXAMPLE 2.4

Differentiate the following function:

$$M = t^4 + 2t^3 \tag{2.13}$$

In this case

$$f(t) = t^4$$

and

$$N(t) = 2t^3$$

Thus, with the aid of equation (2.12), we get

$$\begin{aligned} \frac{dM}{dt} &= \frac{d[t^4]}{dt} + \frac{d[2t^3]}{dt} \\ &= 4t^3 + 6t^2 \end{aligned} \tag{2.14}$$

RULE V

If $M = e^y$, then

$$\frac{dM}{dt} = e^y \frac{dy}{dt}, \tag{2.15}$$

where

Y is a function of t.

EXAMPLE 2.5

Differentiate the following function:

$$M = e^{-\lambda t}, \tag{2.16}$$

where

λ is a constant.

In this case $y = -\lambda t$. Therefore with the aid of equation (2.15), we get

$$\frac{dM}{dt} = e^{-\lambda t} \cdot \frac{d(-\lambda t)}{dt} = e^{-\lambda t}(-\lambda) \tag{2.17}$$

Therefore,

$$\frac{dM}{dt} = -\lambda e^{-\lambda t} \tag{2.18}$$

2.2.2. Integration

Integration is the inverse operation of differentiation and is used to find the area under a curve of a specified function over a given interval.

Integration Rules

Several rules associated with integration are as follows:

RULE I

If $\frac{dF}{dt} = f(t)$, then F is known as the antiderivative of function $f(t)$, or the indefinite integral of $f(t)$. The indefinite integral of $f(t)$ is denoted as follows:

$$\int f(t)\, dt, \tag{2.19}$$

where

the symbol $\int$ is known as the *integral sign.*

EXAMPLE 2.6

Prove that

$$\int t^k \, dt = \frac{t^{k+1}}{k+1} + C, \qquad \text{if} \quad k \neq -1 \tag{2.20}$$

where

k is a constant
C is a integration constant.

With the aid of rule I, after differentiating the right hand term of equation (2.10), we get

$$\frac{d\left(\frac{t^{k+1}}{k+1}\right)}{dt} = \frac{(k+1)}{(k+1)} \cdot t^k$$

$$\frac{d\left(\frac{t^{k+1}}{k+1}\right)}{dt} = t^k \tag{2.21}$$

Thus, from the relationships between equations (2.19)–(2.21), we get

$$\int t^k \, dt = \frac{t^{k+1}}{k+1} + C, \qquad \text{if} \quad k \neq -1 \tag{2.22}$$

RULE II

If the $f(t)$ is continuous over a closed interval $[\alpha, \beta]$ and if F is the integral (antiderivative) of $f(t)$ in the specified interval, then

$$\int_\alpha^\beta f(t) \cdot dt = F(\beta) - F(\alpha) \tag{2.23}$$

EXAMPLE 2.7

Find

$$\int_2^3 t^2 \, dt \tag{2.24}$$

With the aid of equation (2.23),

$$\int_2^3 t^2 \, dt = \left.\frac{t^{2+1}}{(2+1)}\right|_2^3 = \left(\frac{3^3}{3}\right) - \left(\frac{2^3}{3}\right)$$

$$= \frac{19}{3}$$

RULE III

If $f(t) = \lambda e^{-\lambda t}$, then

$$\int_{\alpha}^{\beta} \lambda e^{-\lambda t} \cdot dt = \left. \frac{-\lambda\, e^{-\lambda t}}{\lambda} \right|_{\alpha}^{\beta} = e^{-\lambda\alpha} - e^{-\lambda\beta}, \tag{2.25}$$

where

λ is a constant.

EXAMPLE 2.8

Evaluate

$$\int_{0}^{t} 2e^{-2\tau}\, d\tau \tag{2.26}$$

Thus, with the aid of equation (2.25) we get

$$\int_{0}^{t} 2e^{-2\tau}\, d\tau = e^{-2(0)} - e^{-2(t)} = 1 - e^{-2t} \tag{2.27}$$

2.2.3. The Quadratic Equation

This equation occurs in reliability analysis. The standard form of the equation is as follows:

$$\alpha t^2 + \beta t + \gamma = 0, \qquad \text{for} \quad \alpha \neq 0, \tag{2.28}$$

where

t is the variable
α, β, and γ are the constants.

Roots of the quadratic equation (2.28) are given by

$$t = \frac{-\beta \pm (\beta^2 - 4\alpha\gamma)^{1/2}}{2\alpha} \tag{2.29}$$

If t_1 and t_2 are the roots then

$$t_1 \cdot t_2 = \frac{\gamma}{\alpha} \tag{2.30}$$

and

$$t_1 + t_2 = -\beta/\alpha \tag{2.31}$$

For real α, β, and γ the types of roots can be determined from the following conditions:

1. If $(\beta^2 - 4\alpha\gamma) = 0$, the roots are equal and real.
2. If $(\beta^2 - 4\alpha\gamma) < 0$, the roots are complex conjugate.
3. If $(\beta^2 - 4\alpha\gamma) > 0$, the roots are unequal and real.

EXAMPLE 2.9

Solve

$$t^2 - 2t + 1 = 0$$

with the aid of equation (2.29).

Thus, in this case

$$\alpha = 1, \quad \beta = -2, \quad \text{and} \quad \gamma = 1$$

Therefore, from equation (2.29), we get

$$t = \frac{-(-2) \pm [(-2)^2 - 4 \cdot (1) \cdot 1]^{1/2}}{2} \tag{2.32}$$

In this case $(\beta^2 - 4\alpha\gamma) = 0$; therefore the roots are equal and real. Thus, from equation (2.32) we get the roots

$$t_1 = 1 \quad \text{and} \quad t_2 = 1$$

2.3 BASIC PROBABILITY THEORY

The probability theory is the backbone of reliability and maintainability engineering. It was probably applied initially to engineering systems, with respect to reliability, in the 1930's.[1-4] Four fundamental axioms of probability are as follows:

1. The probability for the outcome of an event X, is a number between zero and one.

$$0 \le P(X) \le 1 \tag{2.33}$$

2. The probability of the union of two mutually exclusive events, for example, X, and the complement of X (i.e. $\bar{X}$) is

$$P(X \cup \bar{X}) = P(X) + P(\bar{X}) = 1, \tag{2.34}$$

where

$\cup$ denotes union.

3. The probability of a specific event, S (known as the entire sample space) is equal to one.

$$P(S) = 1 \tag{2.35}$$

4. The probability of union of mutually exclusive events $X_1, X_2, X_3, \cdots, X_k$ is

$$P(X_1 \cup X_2 \cup X_3 \cdots \cup X_k) = P(X_1) + P(X_2) + P(X_3) + \cdots + P(X_k) \tag{2.36}$$

The probability of success, p_s, is given by

$$p_s = \frac{n_s}{n_{pO}}, \tag{2.37}$$

where

n_s is the number of successes
n_{pO} is the number of possible outcomes.

Similarly, the probability of failure, p_f, is given by

$$p_f = \frac{n_f}{n_{pO}}, \tag{2.38}$$

where

n_f is the number of failures.

Therefore,

$$p_s = \frac{S_w}{S_w + F_w}, \tag{2.39}$$

where

S_w denotes the number of ways success can occur.
F_w denotes the number of ways failure can occur.

and

$$p_f = \frac{F_w}{S_w + F_w} \tag{2.40}$$

Table 2.1. Defective items produced monthly.

Month	Number of Defective Items
January	10
February	5
March	15
April	7
May	9
June	4

EXAMPLE 2.10

For each month of a six month period, the number of defects in a production run were as given in Table 2.1. Each month 500 items were produced. Calculate the probability of an item being defective.

The total number of defects for the six month period

$$= 10 + 5 + 15 + 7 + 9 + 4 = 50$$

Therefore, the average number of defects, D_a, per month are given by

$$D_a = \frac{50}{6} = 8.333 \text{ defects/month}$$

The probability of an item being defective is

$$\frac{D_a}{500} = \frac{8.333}{500} = 0.0167$$

2.3.1. Union of Events

In this case we are concerned with the probability of the union of two events X_1 and X_2. Both events are independent but not mutually exclusive. Thus, the probability of the union of events X_1 and X_2 is given by

$$P(X_1 \cup X_2) = P(X_1) + P(X_2) - P(X_1) \cdot P(X_2) \tag{2.41}$$

where

$P(X_i)$ is the occurrence probability of event X_i, for $i = 1, 2$.
The symbol $\cup$ denotes union.

Similarly the probability of the union of k independent (but not mutually exclusive) events is given by

$$P(X_1 \cup X_2 \cup X_3 \cup \cdots \cup X_k) = 1 - \prod_{i=1}^{k} [1 - P(X_i)] \qquad (2.42)$$

where

$P(X_i)$ is the occurrence probability of event X_i, for $i = 1, 2, 3, \cdots, k$.

EXAMPLE 2.11

A quality control engineer selects two items X_1 and X_2. Compute the probability that item X_1 or item X_2 or both is good if the probabilities of items X_1 and X_2 being good are 0.8 and 0.96, respectively.

By substituting the specified data into equation (2.41) we get

$$P(X_1 \cup X_2) = 0.8 + 0.96 - (0.8) \cdot (0.96)$$
$$= 0.992$$

Thus the probability that item X_1 or item X_2 or both is good is 0.992.

2.3.2. Intersection of Events

The probability of intersection of two independent events, X_1 and X_2, is given by

$$P(X_1 \cap X_2) = P(X_1) \cdot P(X_2) \qquad (2.43)$$

where

the symbol $\cap$ denotes intersection.

Similarly for k independent events, the equation (2.42) is generalized to the following form:

$$P(X_1 \cap X_2 \cap X_3 \cap \cdots \cap X_k) = \prod_{i=1}^{k} [P(X_i)] \qquad (2.44)$$

EXAMPLE 2.12

A person selects three items X_1, X_2, and X_3. Compute the probability of all three items being good if the probabilities of items X_1, X_2, and X_3 being good are 0.9, 0.8, and 0.85 respectively.

Substituting the given data into equation (2.44) leads to

$$P(X_1 \cap X_2 \cap X_3) = P(X_1) \cdot P(X_2) \cdot P(X_3)$$
$$= (0.9) \cdot (0.8) \cdot (0.85)$$
$$= 0.612$$

Thus the probability of all the three items being good is 0.612.

2.3.3. Conditional Probability

Here, we are concerned with calculating the probability of occurrence, $P(Y)$, of an event, Y, where the event Y is dependent upon a number of mutually exclusive events, X_j. Therefore, from reference 5 we get

$$P(Y) = \sum_{j=1}^{k} P(Y/X_i) \cdot P(X_i) \tag{2.45}$$

where

k is the number of mutually exclusive events.
$P(Y/X_i)$ is the conditional probability of event Y occurring, given that event Y_i has occurred.
The symbol "/" denotes GIVEN.

When the event Y occurrence is dependent upon just two mutually exclusive events, the equation (2.45) simplifies to the following form:

$$\begin{aligned} P(Y) &= \sum_{j=1}^{2} P(Y/X_i) \cdot P(X_i) \\ &= P(Y/X_1) \cdot P(X_1) + P(Y/X_2) \cdot P(X_2) \end{aligned} \tag{2.46}$$

For practical application, the success and failure of component X, can be represented by X_1 and X_2, respectively. The equation (2.46) is essentially the same as equation (3.1) in Chapter 3. In Chapter 3, the equation (3.1) was used to develop reliability expressions for series and parallel networks.

2.4. PROBABILITY DISTRIBUTION OF A CONTINUOUS RANDOM VARIABLE

Probability distributions such as exponential, Rayleigh, and Weibull are included in the category of continuous random variable distributions. These distributions are frequently used in reliability analysis work. This section briefly discusses these three distributions along with basis definitions.

2.4.1. Basic Definitions

This section presents five important definitions associated with probability distribution of a continuous random variable.

DEFINITION I

The failure density function is defined as follows:

$$f(t) = z(t) \exp\left[-\int_0^t z(\tau) \cdot d\tau\right] \tag{2.47}$$

where

t is time
$z(t)$ is the hazard rate, sometimes denoted by $\lambda(t)$.

The failure density function can also be obtained from any of the following three relationships:

$$f(t) = \frac{dF(t)}{dt} \tag{2.48}$$

$$f(t) = \frac{-dR(t)}{dt} \tag{2.49}$$

and

$$f(t) = z(t)[1 - F(t)] \tag{2.50}$$

where

$F(t)$ is the cumulative distribution function
$R(t)$ is the reliability function.

DEFINITION II

The cumulative distribution function, $F(t)$, of a continuous random variable is given by

$$F(t) = \int_0^t f(\tau) \cdot d\tau \tag{2.51}$$

At $t = 0$, $F(0) = 0$ and at $t = \infty$, $F(\infty) = 1$

DEFINITION III

The reliability function is given by any of the following relationships:

$$R(t) = 1 - F(t) \tag{2.52}$$

$$R(t) = \int_t^\infty f(\tau) \cdot d\tau \tag{2.53}$$

and

$$R(t) = \exp\left[-\int_0^t z(\tau) \cdot d\tau\right] \tag{2.54}$$

DEFINITION IV

The mean value, μ, of a distribution is given by

$$\mu = \int_0^\infty tf(t) \cdot dt \tag{2.55}$$

DEFINITION V

The variance, σ^2, is defined by

$$\sigma^2 = \int_0^\infty (t - \mu)^2 f(t)\, dt \tag{2.56}$$

2.4.2. Exponential Distribution

Exponential distribution is probably the most widely used distribution in reliability engineering. The distribution failure density function, the cumulative distribution function, the reliability function, the mean value, and the variance are given by

$$f(t) = \lambda\, e^{-\lambda t}\ \lambda > 0, \qquad t \geq 0 \tag{2.57}$$

where

λ is the scale parameter. In reliability work it is known as the constant failure rate of an item

t is time.

$$F(t) = 1 - e^{-\lambda t} \tag{2.58}$$

$$R(t) = e^{-\lambda t} \tag{2.59}$$

$$\mu = \frac{1}{\lambda} \tag{2.60}$$

$$\sigma^2 = \frac{1}{\lambda^2} \tag{2.61}$$

EXAMPLE 2.13

An engineering component failure times are defined by

$$f(t) = \lambda\, e^{-\lambda t} = (0.007)\, e^{-(0.007)t} \tag{2.62}$$

where

λ = 0.007 failures/hour. In reliability engineering, the parameter λ is known as the constant failure rate.

Compute the component probability of failure for a 60 hour mission.

Substituting equation (2.62) into equation (2.51) and integrating, results in

$$\begin{aligned} F(t) &= \int_0^t (0.007) e^{-(0.007)\tau} \cdot d\tau \\ &= 1 - e^{-(0.007)t} \end{aligned} \tag{2.63}$$

Since $t = 60$ hours from the above equation, we get:

$$F(60) = 1 - e^{-(0.007)(60)}$$
$$= 0.3429$$

The component probability of the failure for the specified mission time is 0.3429.

2.4.3. Rayleigh Distribution

Rayleigh distribution is also used in reliability engineering. The distribution failure density function, the cumulative distribution function, the reliability function, the mean value, and the variance are given by

$$f(t) = \theta t \exp\left[-\frac{\theta t^2}{2}\right], \qquad \text{for} \quad t \geq 0 \tag{2.64}$$

where

θ is the parameter
t is time.

$$F(t) = 1 - \exp\left[-\frac{\theta t^2}{2}\right] \tag{2.65}$$

$$R(t) = \exp\left[-\frac{\theta t^2}{2}\right] \tag{2.66}$$

$$\mu = \left(\frac{\pi}{2\theta}\right)^{1/2} \tag{2.67}$$

$$\theta^2 = \frac{2}{\theta}\left(1 - \frac{\pi^2}{4}\right) \tag{2.68}$$

2.4.4. Weibull Distribution

Weibull distribution was developed by Weibull and plays an important role in the statistical analysis of failure data because of its flexibility.[6] Exponential and Rayleigh distributions are special cases of the Weibull distribution. The distribution failure density function, the cumulative distribution function, the reliability function, the mean value, and the variance are given by

$$f(t) = \frac{\theta t^{\theta-1}}{\beta^\theta} \exp\left[-\left(\frac{t}{\beta}\right)^\theta\right], \qquad \text{for} \quad t \geq 0, \quad \beta > 0, \quad \theta > 0 \tag{2.69}$$

where

θ is the shape parameter
β is the scale parameter
t is time.

$$F(t) = 1 - \exp\left[-\left(\frac{t}{\beta}\right)^{\theta}\right] \tag{2.70}$$

$$R(t) = \exp\left[-\left(\frac{t}{\beta}\right)^{\theta}\right] \tag{2.71}$$

$$\mu = \beta\Gamma\left(\frac{1+\theta}{\theta}\right) \tag{2.72}$$

$$\theta^2 = \beta^2\left[\Gamma\left(1+\frac{2}{\theta}\right) - \Gamma^2\left(1+\frac{1}{\theta}\right)\right] \tag{2.73}$$

2.5. SUMMARY

This chapter presents some basic concepts of mathematics that are necessary to understand subsequent chapters of this book. Differentiation and integration are briefly discussed with the aid of solved examples. A formula to find roots of the quadratic equation is presented. Four fundamental axioms of probability are discussed. Equation for the probability of the union of k events, and equation for the probability of the intersection of k independent events are presented. Conditional probability is briefly discussed. Definitions of failure density function, cumulative distribution function, reliability function, mean value, and variance are presented. Finally, the exponential, Rayleigh, and Weibull distributions are discussed.

2.6. EXERCISES

1. If $y = 20t^2 + t$, find $\frac{dy}{dt}$

2. Prove that

$$\frac{d[\ln t]}{dt} = \frac{1}{t}$$

3. If $y = 1 - \exp\left[-\left(\frac{t}{\beta}\right)^{\theta}\right]$, find $\frac{dy}{dt}$

where

β and θ are constants.

4. Evaluate

$$\int_0^3 (t^2 + t + 1)\, dt$$

5. Differentiate

$$y = \frac{e^t - 1}{e^t + 3}$$

where

t is the variable.

6. Integrate

$$\int_0^\infty \exp\left[-\left(\frac{t}{\beta}\right)^\theta\right] dt$$

where

t is the variable
θ, β are the constants.

7. Evaluate

$$\int_0^\infty \lambda e^{-\lambda t}\, dt$$

where

λ is the parameter.

8. With the aid of equation (2.46), develop an expression for the reliability of the independent component network shown in Figure 2.1. R_i denotes the ith network unit reliability for $i = 1, 2, 3$,

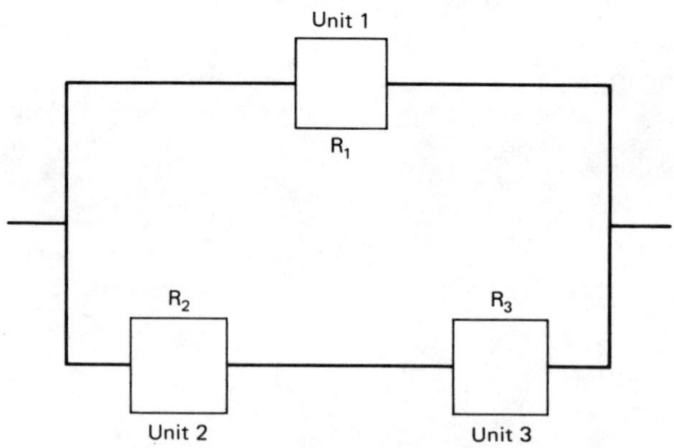

Figure 2.1. A three unit reliability network.

9. Prove that the variance associated with the Weibull distribution is given by

$$\theta^2 = \beta^2 \left[\Gamma\left(1 + \frac{2}{\theta}\right) - \Gamma^2\left(1 + \frac{1}{\theta}\right)\right]$$

where

β is the scale parameter
θ is the shape parameter.

10. Prove that the mean value, μ, of the Rayleigh distribution is given by

$$\mu = \left(\frac{\pi}{2\lambda}\right)^{1/2}$$

where

λ is the parameter.

2.7. REFERENCES

1. Smith, S. A. "Service Reliability Measured by Probabilities of Outage." *Electrical World,* Vol. 103, 1934, pp. 371–374.
2. Benner, P. E. "The Use of the Theory of the Probability to Determine Spare Capacity." *General Electric Review,* Vol. 37, 1934, pp. 345–348.
3. Lyman, W. J. "Fundamental Consideration in Preparing a Master System Plan." *Electrical World,* Vol. 101, 1933, pp. 778–792.
4. Smith, S. A. "Spare Capacity Fixed by Probabilities of Outage." *Electrical World,* Vol. 103, 1934, pp. 222–225.
5. Ang, A. H-S. *Probability Concepts in Engineering Planning and Design.* John Wiley & Sons, Inc., New York, 1975, pp. 56–59.
6. Weibull, W. "A Statistical Distribution Function of Wide Applicability." *J. Appl. Mech.*, Vol. 18, 1951, pp. 293–297.

CHAPTER 3

Introduction to Reliability and Maintainability

3.1. INTRODUCTION

Recently, the importance of reliability and maintainability in the management field is growing due to factors such as system complexity, system sophistication, use of previously untried technology, and so on. Although reliability and maintainability are relatively new fields, the application of probability concepts to problems associated with electric power generation, goes as far back as the 1930's. A number of papers were published on this topic in 1930's.[1-6] But the real advent of the reliability concept is considered to be World War II when the Germans applied this concept to their V1 and V2 rockets.

During the War, the United States armed forces also felt a pressing need for reliable equipment. Thus, at the end of the war various military reliability-related studies were initiated, the results of which were startling. For example, one of the studies revealed that between two-thirds and three-fourths of United States Army hardware (i.e. equipment) was either undergoing repairs or out of commission. As a result of such studies, an ad hoc committee on reliability was established in 1950 by the United States Department of Defense. Two years later, this committee became a permanent group known as the Advisory Group on the Reliability of Electronic Equipment (AGREE). A report prepared by this group appeared in July of 1957, and since then, extensive literature on the subject has appeared. Today, terms such as reliability engineering department, reliability engineer, and reliability engineering manager are well-known in many industries.

As a result of the awareness of the reliability problem during and after World War II, equipment maintainability has also been receiving a significant amount of attention. In June of 1959, a military document, MIL-M-26512, was produced by the United States Air Force. The document was the result of a growing concern for maintainability, and was followed by other military published maintainability documents, such as MIL-STD-778 (definitions of maintainability terms), MIL-STD-470 (maintainability program requirements), MIL-STD-471 (maintainability demonstration), MIL-STD-472 (maintainability prediction), MIL-STD-721B (definition of effectiveness terms for reliability, maintainability, human factors, and safety), AMCP 706-134 (maintainability guide for design), and NAVSHIPS 94324 (maintainability design criteria for designers of shipboard electronic equipment). In addition, some books and a large number of articles have also appeared on the subject of maintainability.

This chapter discusses several aspects of reliability as well as the introductory aspect of maintainability.

3.2. SELECTIVE DEFINITIONS AND SOURCES OF CURRENT RELIABILITY AND MAINTAINABILITY INFORMATION

This section presents the definitions of selective terms used in reliability and maintainability engineering, and the sources of current reliability and maintainability information.

Some frequently used terms in the reliability and maintainability field are defined below:[7,8]

- *Reliability*—the probability that a unit will function normally when used according to specified conditions for at least a stated period of time.
- *Maintainability*—the probability that a failed unit is put back to satisfactory, operable condition in a given down time.
- *Redundancy*—the existence of two or more means for carrying out a stated function.
- *Active Redundancy*—a redundancy whereby the total number of redundant units are functioning simultaneously rather than being switched on at the moment of need.
- *Standby redundancy*—a redundancy whereby the alternative means of carrying out the function is inactive until required and is switched on when the primary means of performing the function fails.
- *Failure Rate*—the failures of a product per unit measure of life. Examples of unit measure of life are time, miles, and cycles.
- *Corrective Maintenance*—the actions carried out (because of failure) to repair a component, equipment, or a system to a stated condition.
- *Mean-Time-to-Repair*—the total corrective maintenance time over the total corrective maintenance actions during a specified time period.
- *Downtime*—the component of time during which the unit is unable (i.e. not in condition) to carry out its function.

In order to obtain current information on reliability and maintainability, the following are the main sources:

94 *Journals*

1. Microelectronics and Reliability—an International Journal, published by the Pergamon Press, Oxford, England, since the early 1960's.
2. IEEE Transactions on Reliability—jointly published by the American Society for Quality Control (ASQC) and IEEE, USA, since the 1950's.
3. Reliability Review—a News and Opinion quarterly, published by the reliability division of the American Society for Quality Control, USA, since 1981.
4. International Journal of Quality and Reliability Management—published by MCB University Press Limited, Bradford, England, since 1984.
5. Quality and Reliability Engineering International—published by John Wiley & Sons Ltd., Chichester, England, since 1985.
6. Maintenance Management International—published by Elsevier Scientific Publishing Company, Amsterdam, The Netherlands, since 1979. (Frequently publishes articles on reliability and maintainability.)
7. IEEE Transactions on Power Apparatus and Systems—published by IEEE, USA. (Frequently publishes articles on power system reliability and maintainability.)

Conference Proceedings

1. Proceedings of the Annual Reliability and Maintainability Symposium—published by IEEE. The symposium is held annually in the United States.
2. Proceedings of the Annual International Reliability, Availability, and Maintainability Conference for the Electric Power Industry—published by IEEE. The Conference is held annually usually in United States.
3. Proceedings of the International Conference on Reliability and Maintainability (being held after every two years in France).
4. Proceedings of the International Conference on Reliability and Exploitation of Computer Systems (being held after every two years in Poland).

3.3. LIFE CHARACTERISTIC CURVE

This section describes the well known bathtub hazard rate curve of a product shown in Figure 3.1. Because the failure rate of a product changes with time, in many cases, the curve shown in Figure 3.1 is used to represent the product's hazard rate. This curve has three distinct regions known as the "burn-in region," the "useful life region," and the "wearout region."

As shown in Figure 3.1 the region between interval $[0, t_a]$ is known as the

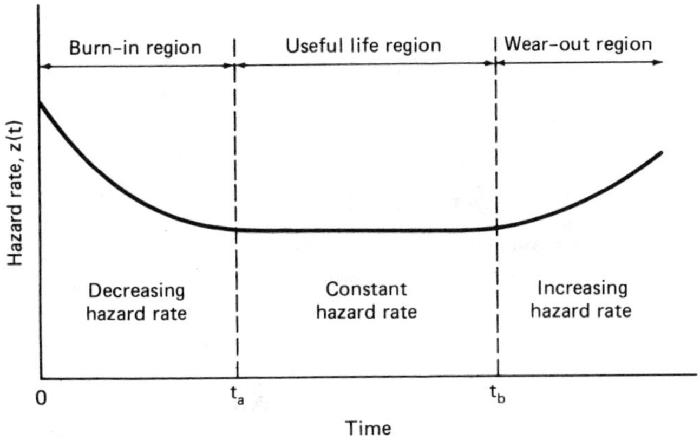

Figure 3.1. Bathtub hazard rate curve.

"burn-in region," the "infant mortality region," the "debugging region," or the "break-in region."

Many of the reasons for the decreasing hazard rate region failures are listed in Table 3.1.[9] Most producers of manufactured engineering equipment provide a "burn-in period" for their products.

In the useful life region of Figure 3.1, the hazard rate is constant. In addition, the failures occur unpredictably and therefore, are known as random failures. Some of the reasons for these failures are given in Table 3.2.[9]

In the "wear-out region" of Figure 3.1, the hazard rate starts increasing from time t_b. This increase indicates the end of the useful life period of the product in question. Some causes for the failures during the "wear-out pe-

Table 3.1. Reasons for "burn-in region" failures.

No.	Reason
1	Inadequate quality control.
2	Inadequate manufacturing methods.
3	Substandard materials and workmanship.
4	Wrong start-up and installation.
5	Difficulties because of assembly.
6	Inadequate debugging.
7	Inadequate processes and human error.
8	Inadequate handling methods and wrong packaging.

Table 3.2. Reasons for "useful life region" failures.

No.	Reason
1	Causes which cannot be explained.
2	Human errors, abuse, and natural failures.
3	Unavoidable failures: these cannot be avoided by even the most effective preventive maintenance practices.
4	Undetectable defects.
5	Low safety factors.
6	Higher random stress than expected.

Table 3.3. Causes for "wear-out region" failures.

No.	Reason
1	Inadequate maintenance
2	Wear due to friction
3	Wear due to aging
4	Wrong overhaul practices
5	Corrosion and creep
6	Designed-in life of the product is short.

riod" are listed in Table 3.3.[9] Failures occurring in the "wear-out period" can be reduced with properly designed replacement and preventive maintenance policies and procedures.

3.4. RELIABILITY NETWORKS

This section is concerned with the reliability evaluation of commonly known networks such as series, parallel, r-out-of-m, etc. It is assumed throughout this section that each network component fails independently. In addition, it is assumed that each component of the network can only be in either of two states (i.e. good or failed). In all the networks, the reliability of each unit is assumed to be constant.

3.4.1. Series Network

A two unit series network is shown in Figure 3.2. This network represents a two independent unit system whose components (units) are connected in

Figure 3.2. A two unit series system.

series. It means if any one of the components fails, the entire system fails. In other words both components must operate successfully for the system success. A typical example of such a system is an aircraft whose two independent engines are needed for the aircraft to fly successfully. In order to obtain an expression for reliability of network shown in Figure 3.2., we make use of the following equation:

$$R_s = R_c R_{sg} \text{ (system good if the critical component } C \text{ is good)} + (1 - R_c) R_{sgb} \text{ (system good if the critical component } C \text{ is bad)} \quad (3.1)$$

where

R_s	is the system reliability
R_c	is the critical component C reliability
$(1 - R_c)$	is the critical component C unreliability
R_{sg} (system good if the critical component C is good)	is the system reliability if the probability of C being good is 100%
R_{sgb} (system good if the critical component C is bad)	is the system reliability if the probability of C being good is 0%

Equation (3.1) can be used to obtain a reliability expression for the series system shown in Figure 3.2, by labeling either unit 1 or unit 2 of the diagram as the critical element C. For our purpose we labelled unit 1 as the critical element as shown in Figure 3.3. When unit 1 is 100% reliable, the Figure 3.3 diagram reduces to the diagram shown in Figure 3.4. This Figure shows only unit 2 because unit 1 is 100% reliable (i.e., it never fails). Thus, the series system reliability, R_{sg}, from Figure 3.4 with 100% reliable critical element C is given by

$$R_{sg} = 1 \cdot R_2 = R_2 \quad (3.2)$$

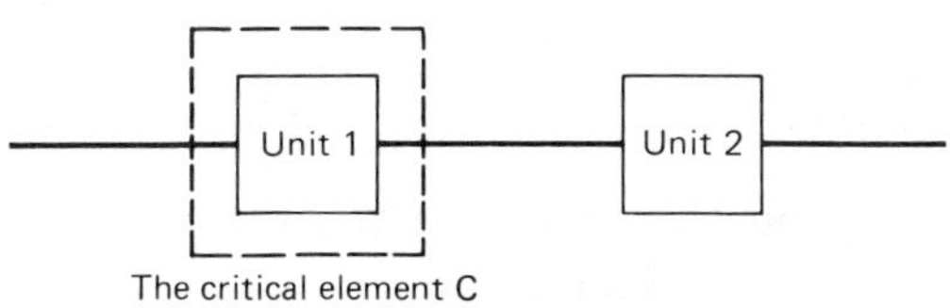

Figure 3.3. Series system with identified critical element C.

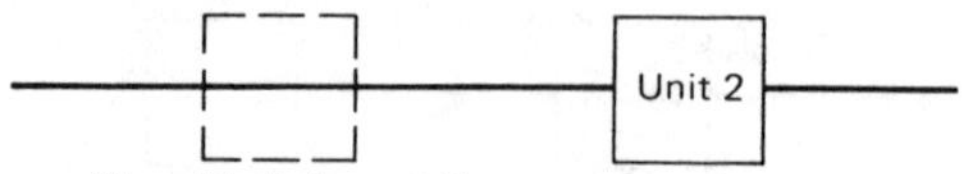

Figure 3.4. Series system with 100% reliable critical element C (i.e., unit 1).

Figure 3.5. Series system with 0% reliable unit 1 (i.e., critical element C).

where

R_2 is the unit 2 reliability.

Similarly, when the probability of the critical element C being good is 0%, Figure 3.3 reduces to the one shown in Figure 3.5.

Thus, from Figure 3.5 with 0% reliable critical element C we get

$$R_{sgb} = 0 \cdot R_2 = 0 \tag{3.3}$$

Substituting equations (3.2) and (3.3) into equation (3.1) results in

$$R_s = R_c R_2 + (1 - R_c) \cdot 0 \tag{3.4}$$

Since $R_c = R_1$, equation (3.4) reduces to

$$R_s = R_{ss} = R_1 \cdot R_2 \tag{3.5}$$

where

R_{ss} is the series system reliability.
R_1 is the unit 1 (also known as the critical element C) reliability.

Similarly, for n independent and non-identical unit series system, equation (3.5) is generalized to the following form:

$$R_{ss} = R_1 \cdot R_2 \cdot R_3 \cdot \cdots R_n \tag{3.6}$$

where

R_i is the ith unit reliability, for $i = 1, 2, 3, 4, \cdots, n$.

For identical components, equation (3.6) simplifies to

$$R_{ss} = R^n \tag{3.7}$$

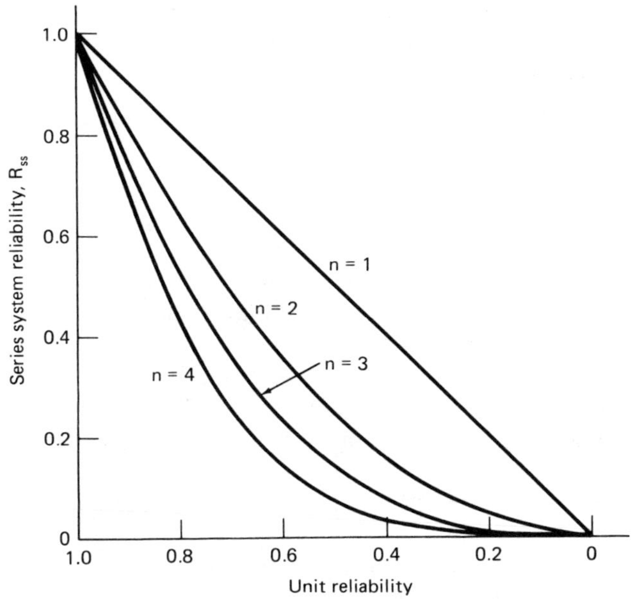

Figure 3.6. Identical unit series system reliability plots.

where

R is each unit's reliability.

The plots of equation (3.7) are shown in Figure 3.6. These plots show the effect that an increasing number of units in series has on system reliability; that is, system reliability decreases as the number of units in series becomes larger.

EXAMPLE 3.1

An automobile has four independent and identical tires. If any of these tires flattens then the vehicle cannot be driven. Calculate the probability of the automobile failing to operate due to a flat tire. Also, if the reliability of each tire is 0.9, what is the reliability of the automobile with respect to tires.

This example represents a four independent unit series system. (The vehicle tires are the units of the series system.) Since all four tires are identical, from equation (3.7) we get

$$R_{ss} = R^4 \tag{3.8}$$

Thus, substituting the specified data into equation (3.8), we get

$$R_{ss} = R^4 = (0.9)^4 = 0.6561$$

Thus, the reliability of the automobile with respect to tires is 0.6561.

Similarly, the probability, Q_{ss}, of the automobile failing to operate due to a flat tire is given by

$$Q_{ss} = 1 - R^4 = 1 - (0.6561) = 0.3439$$

3.4.2. Parallel Network

A two unit, parallel network is shown in Figure 3.7. This network represents a two, independent unit system whose active components (units) are connected in parallel. This means that if any one of the components fails, the system would still operate successfully. In other words, at least one component is needed for the system success. A typical example of such a system is an aircraft whose two independent active engines, enable it to fly normally when at least one engine is functioning successfully. In order to obtain an expression for reliability of configuration shown in Figure 3.7, we make use of the equation (3.1). We select unit 1 as our critical element C as shown in Figure 3.8.

When the critical element is 100% reliable, the Figure 3.8 diagram becomes the one shown in Figure 3.9. Figure 3.9 indicates that the parallel system will never fail.

Thus the parallel system reliability, R_{sg}, from Figure 3.9, with 100% reliable critical element C, is given by

$$R_{sg} = 1 - (1 - R_2)(1 - R_1) = 1 - (1 - R_2)(1 - 1) = 1 \quad (3.9)$$

where

R_1 is the reliability of the parallel unit 1 (i.e. the critical element)
R_2 is the reliability of the parallel unit 2.

In equation (3.9), the reliability of the critical element C (i.e., unit 1) is equal to unity because it is assumed that the critical element is 100% reliable. Similarly, when the probability of the critical element failure shown in Fig-

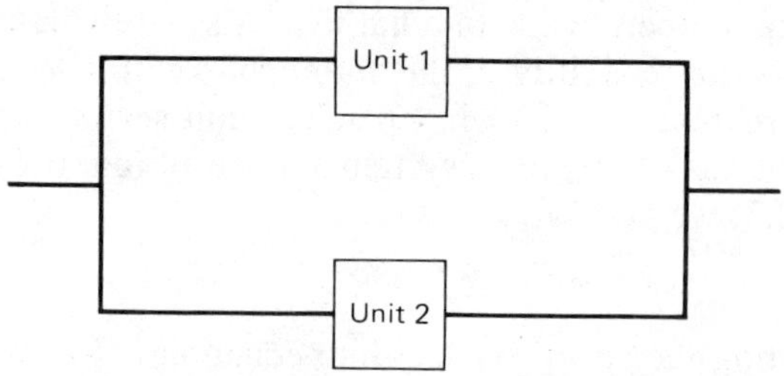

Figure 3.7. A two unit parallel network.

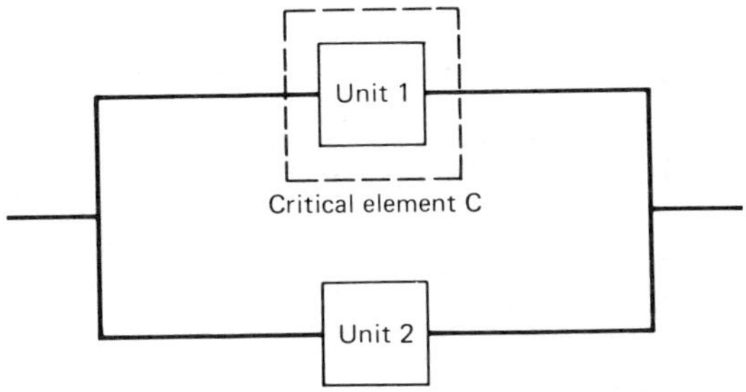

Figure 3.8. Parallel network with identified critical element *C*.

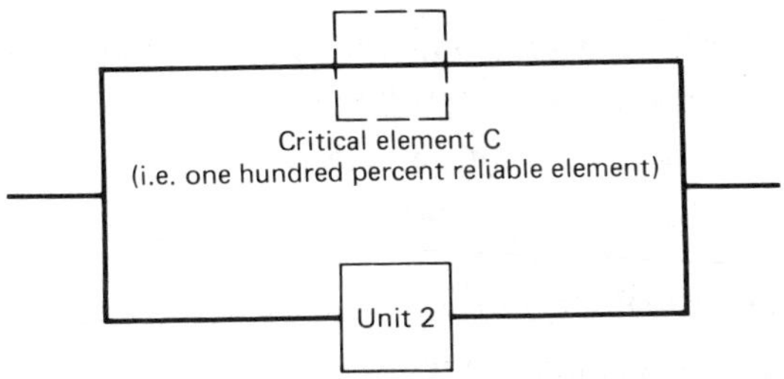

Figure 3.9. Parallel system with 100% reliable critical element *C*.

ure 3.8 is equal to unity (i.e. the reliability of this element is equal to zero), the Figure 3.8 diagram reduces to the diagram shown in Figure 3.10.

The parallel system reliability, R_{sgb}, from Figure 3.10 is given by

$$R_{sgb} = 1 - (1 - R_2)(1 - R_1) = 1 - (1 - R_2)(1 - 0) = R_2 \quad (3.10)$$

In equation (3.10), the reliability of the critical element C (i.e. unit 1) is equal to zero.

By substituting equations (3.9)–(3.10) into equation (3.1), we get

$$R_p = R_s = R_c \cdot 1 + (1 - R_c)R_2 \quad (3.11)$$

where

R_p is the parallel system reliability.

Since $R_c = R_1$, equation (3.11) becomes

$$R_p = R_1 + R_2 - R_1R_2 \quad (3.12)$$

Adding and subtracting unity in the righthand term of equation (3.12) results in

$$\begin{aligned} R_p &= 1 - 1 + R_1 + R_2 - R_1R_2 \\ &= 1 - (1 - R_1)(1 - R_2) \end{aligned} \quad (3.13)$$

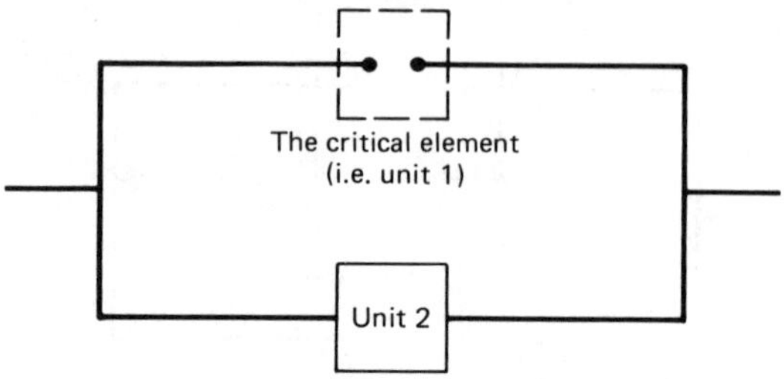

Figure 3.10. The parallel system with 0% reliable unit 1 (i.e., the critical element C).

Similarly, for m independent and non-identical unit parallel system, equation (3.13) is generalized to the following form:

$$R_p = 1 - (1 - R_1)(1 - R_2) \cdots (1 - R_m) \tag{3.14}$$

where

R_i is the ith unit reliability; for $i = 1, 2, 3, \cdots, m$.

For identical units, equation (3.14) reduces to

$$R_p = 1 - (1 - R)^m \tag{3.15}$$

where

R is the reliability of each unit.

The plots of equation (3.15) are shown in Figure 3.11. These plots clearly depict the effect of redundant units on system reliability.

EXAMPLE 3.2

An aircraft has three identical, independent engines that function simultaneously. At least one engine must operate successfully for the aircraft to fly successfully. If the reliability of each engine is 0.8, calculate the probability of the aircraft flying successfully. In this example the values of m and R are 3 and 0.8, respectively. Thus, substituting the given data into equation (3.15) yields

$$R_p = 1 - (1 - 0.8)^3 = 0.992$$

The reliability or probability of the aircraft flying successfully is 0.992.

3.4.3. Complex Network

In this case the independent unit complex network is composed of combinations of earlier series and parallel networks. An example of such a network is shown in Figure 3.12.

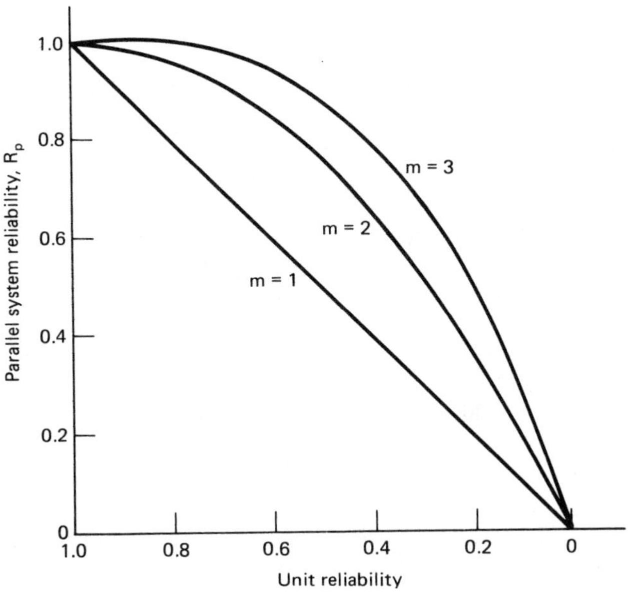

Figure 3.11. Identical unit parallel system reliability plots.

The reliability expression of this type of network can be developed by application of equations (3.6), (3.7), (3.14), and (3.15). For example in Figure 3.12, the subsystems *A*, *B*, and *C* are series, parallel, and parallel networks, respectively. Furthermore, the subsystems *B*, *C*, and *D* are connected in series.

With the aid of equation (3.6), the reliability, R_A, of series subsystem *A* shown in Figure 3.12, is given by

$$R_A = R_3 \cdot R_4 \tag{3.16}$$

where

R_3 is the unit 3 reliability
R_4 is the unit 4 reliability.

With the aid of equation (3.14), the reliability, R_C, of subsystem *C*, shown in Figure 3.12, is given by

$$R_C = 1 - (1 - R_5)(1 - R_A) \tag{3.17}$$

where

R_5 is the unit 5 reliability.

Substituting equation (3.16) into equation (3.17) results in

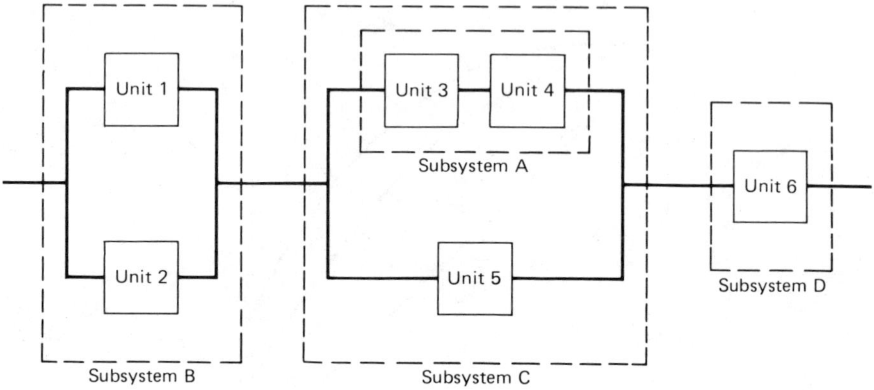

Figure 3.12. A complex network.

$$R_C = 1 - (1 - R_5)(1 - R_3R_4) \tag{3.18}$$

Similarly, the reliability, R_B, of subsystem B, as shown in Figure 3.12, is obtained by utilizing equation (3.14) as follows:

$$R_B = 1 - (1 - R_1)(1 - R_2) \tag{3.19}$$

where

R_1 is the unit 1 reliability
R_2 is the unit 2 reliability.

As shown in Figure 3.12, the subsystems B, C and D are connected in series. Therefore, using equation (3.6), the complex system overall reliability, R_{com}, is

$$R_{com} = R_B \cdot R_C \cdot R_D \tag{3.20}$$

where

$R_D = R_6$, the unit 6 reliability.

Substituting equations (3.18) and (3.19) into equation (3.20) results in

$$R_{com} = [1 - (1 - R_1)(1 - R_2)] \cdot [1 - (1 - R_5)(1 - R_3 \cdot R_4)] \cdot R_6 \tag{3.21}$$

By following the same approach, the reliability of any complex network composed of series and parallel combinations can be evaluated.

EXAMPLE 3.3.

A room contains three identical, independent light bulbs controlled from a single switch, as represented in Figure 3.13. The bulbs and switch fail in-

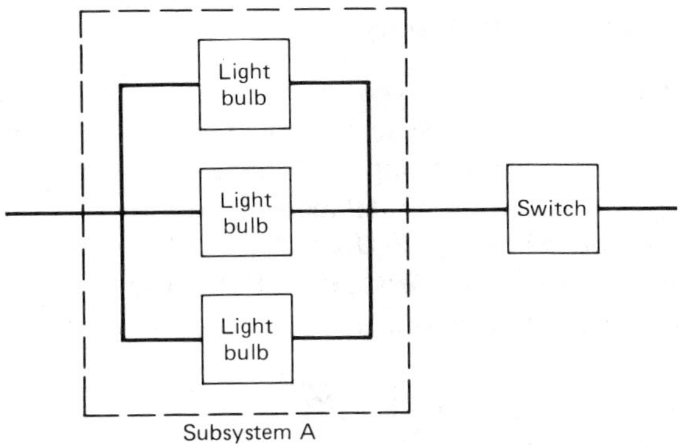

Figure 3.13. The reliability network.

dependently. In addition, the switch can fail to close but can never fail to open. The reliability of the switch and the light bulb are 0.9 and 0.7, respectively. Calculate the probability that the room will be lit.

Using equation (3.15), the reliability of the subsystem A, in Figure 3.11, is

$$R_A = 1 - (1 - R)^3 \tag{3.22}$$

where

R_A is the subsystem A reliability
R is the light bulb reliability.

In Figure 3.13, the subsystem A and the switch form a series system. Thus, with the aid of equation (3.6), we get

$$R_{rl} = R_A \cdot R_s \tag{3.23}$$

where

R_s is the switch reliability.

Substituting equation (3.22) into equation (3.23) results in

$$R_{rl} = [1 - (1 - R)^3] \cdot R_s \tag{3.24}$$

By substituting the specified data for R and R_s into equation (3.24) yields

$$R_{rl} = [1 - (1 - 0.7)^3] \cdot (0.9)$$
$$= 0.8757$$

The probability that the room will be lit is 0.8757.

3.4.4 Partial Active Redundancy

This type of redundancy is also known as the *r*-out-of-*m* unit network. In this case, at least *r* units out-of-*m* active units must function successfully for the system success. A typical example of such a situation is an aircraft with three active engines operating simultaneously. At least two engines must function normally for the aircraft to fly successfully.

If each of the *m* units is independent and identical, the reliability of the system, R_{rm}, with the aid of binomial distribution is given by

$$R_{rm} = \sum_{i=r}^{m} \binom{m}{i} R^i (1 - R)^{m-i} \tag{3.25}$$

where

$$\binom{m}{i} = \frac{m!}{i!(m - i)!}$$

$m! = m(m - 1)(m - 2) \cdots 3.2.1$
R is the unit reliability.

It is to be noted that at $i = m$, equation (3.25) reduces to the equation for the identical unit series network (i.e. equation (3.7)):

$$\begin{aligned} R_{mm} = R_{ss} &= \sum_{i=m}^{m} \binom{m}{i} R^i (1 - R)^{m-i} \\ &= \binom{m}{m} R^m (1 - R)^0 = R^m, \qquad \text{for} \quad m = n \end{aligned} \tag{3.26}$$

Similarly at $i = 1$, equation (3.25) reduces to the equation for the identical unit parallel network (i.e. equation (3.15)):

$$\begin{aligned} R_{1m} = R_p &= \sum_{i=1}^{m} \binom{m}{i} R^i (1 - R)^{m-i} \\ &= 1 - (1 - R)^m \end{aligned} \tag{3.27}$$

2-out-of-3 Unit Network

This is the special case of the *r*-out-of-*m* unit network, at $r = 2$ and $m = 3$. Thus, from equation (3.25), we get 2-out-of-3 unit network reliability

$$\begin{aligned} R_{23} &= \sum_{i=2}^{3} \binom{3}{i} R^i (1 - R)^{3-i} \\ &= \binom{3}{2} R^2 (1 - R) + \binom{3}{3} R^3 (1 - R)^0 \\ &= 3R_2^2 - 2R^3 \end{aligned} \tag{3.28}$$

To compare the reliability of the 2-out-of-3 unit network with that of the single unit network, we must find the crossover point by equating each network's reliability as follows:

[Reliability of the 2-out-of-3 unit system] = [Reliability of single unit system]

$$3R^2 - 2R^3 = R \tag{3.29}$$

Dividing both sides of equation (3.29) by R we get

$$3R - 2R^2 = 1 \tag{3.30}$$

Therefore,

$$2R^2 - 3R + 1 = 0 \tag{3.31}$$

The above equation is the quadratic equation. Thus, the roots of equation (3.31) are given by

$$R_1, R_2 = \frac{3 \pm \sqrt{9 - 4.2.1}}{2 \times 2} \tag{3.32}$$

From equation (3.32) the roots are

$$R_1, = 1, \quad R_2 = \tfrac{1}{2} \quad \text{or} \quad R = 1, \tfrac{1}{2}$$

In other words, the reliability of a single unit system is equal to that of the 2-out-of-3 unit system at $R = 1$ or $\frac{1}{2}$. This fact is shown in Figure 3.14.

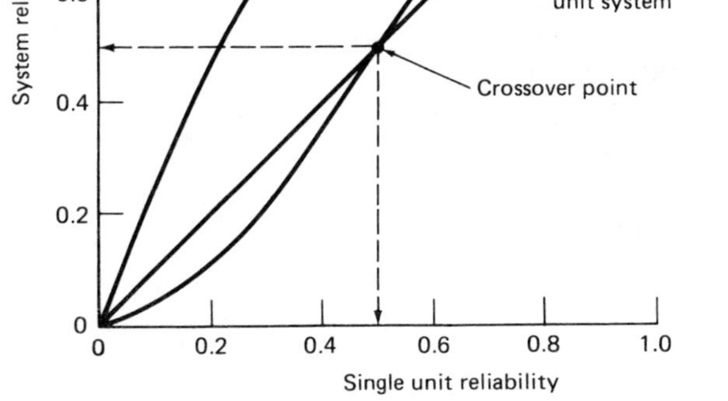

Figure 3.14. Reliability plots of *r*-out-of-*m* unit system.

The 2-out-of-3 unit system reliability is greater than the reliability of a single unit system when the reliability of the single unit is between 0.5 and 1. As shown in Figure 3.14, in comparison to the 1-out-of-3 unit system (i.e. the parallel system), the 2-out-of-3 unit system reliability is much less.

EXAMPLE 3.4

An aircraft has three independent, active, and identical engines whose probability of success is 0.9. At least two engines must operate successfully for the aircraft to fly successfully. Compute the reliability of the aircraft flying successfully.

This situation can be represented by the 2-out-of-3 unit configuration. Thus, from equation (3.28) the reliability R_{air}, of the aircraft is given by

$$\begin{aligned} R_{air} &= 3R^2 - 2R^3 \\ &= 3(0.9)^2 - 2(0.9)^3 \\ &= 2.43 - 1.4580 \\ &= 0.9720 \end{aligned}$$

The probability of the aircraft flying successfully is 97.20%.

3.4.5. Other Configurations

This section presents the reliability expressions for independent and active unit reliability networks not previously discussed. The reliability expressions for the miscellaneous networks are presented in Table 3.4. These expressions were developed by following the procedure used for previous networks.

3.5. RELIABILITY AND MAINTAINABILITY MEASURES

This section presents general formulas for obtaining item reliability, mean time to failure (MTTF), hazard rate, maintainability function, and steady state availability and unavailability.

3.5.1. Reliability Functiion

The reliability, $R(t)$, of an item is given by[10]

$$R(t) = \exp\left[-\int_0^t \lambda(\tau)\, d\tau\right] \qquad (3.33)$$

where

t is time
$\lambda(t)$ is the item hazard rate.

The following equation is also used to compute the reliability of an item

$$\begin{aligned} R(t) &= 1 - F(t) \\ &= 1 - \int_0^t f(\tau)\, d\tau \qquad (3.34) \\ &= \int_t^\infty f(\tau)\, d\tau \end{aligned}$$

where

$f(t)$ is the failure density function
$F(t)$ is the cumulative distribution function.

Table 3.5 presents the reliability functions associated with exponential, Rayleigh, and Weibull distributions. These reliability expressions were developed with the aid of equation (3.34).

EXAMPLE 3.5

The hazard rate of an electronic item is defined by

$$\lambda(t) = \lambda \qquad (3.35)$$

Obtain an expression for the item's reliability function, and calculate its reliability for a 75 hour mission if the value of the item failure rate, λ, is 0.009 failures/hour.

Substituting equation (3.35) into equation (3.33) yields

$$R(t) = \exp\left[-\int_0^t \lambda\, d\tau\right] = e^{-\lambda t} \qquad (3.36)$$

Plugging the specified data into equation (3.36), results in

$$R(75) = e^{-[0.009][75]} = 0.5092$$

Hence, the reliability of the electronic item for the specified mission time is 0.5092.

EXAMPLE 3.6

The failure density function of a component is as follows:

$$f(t) = \lambda e^{-\lambda t} \qquad (3.37)$$

Table 3.4. Reliability expressions for selective configurations. R_n denotes network reliability. R_i denotes ith unit reliability for $i = 1, 2, 3, 4, 5$.

Reliability Networks	Network Reliability Equation (non-identical units)	Network Reliability Equation (identical units)
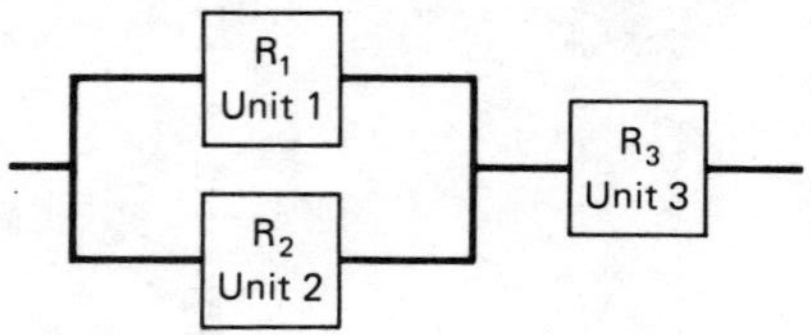	$R_n = [1 - (1 - R_1)(1 - R_2)]R_3$	$R_n = R[1 - (1 - R)^2]$ where $R = R_1 = R_2 = R_3$
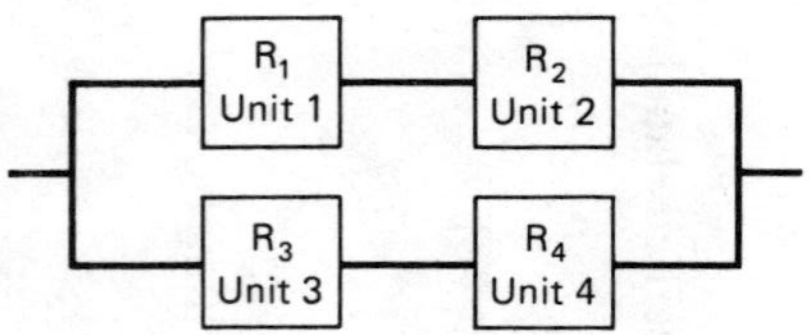	$R_n = 1 - (1 - R_1R_2)(1 - R_3R_4)$	$R_n = 1 - (1 - R^2)^2$ where $R = R_1 = R_2 = R_3 = R_4$
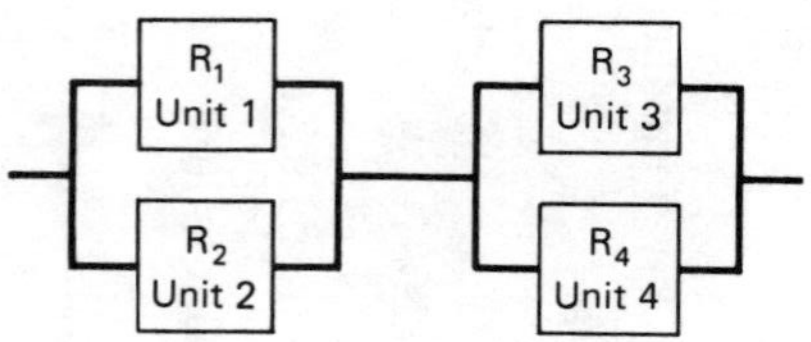	$R_n = [1 - (1 - R_1)(1 - R_2)]$ $[1 - (1 - R_3)(1 - R_4)]$	$R_n = [1 - (1 - R)^2]^2$ where $R = R_1 = R_2 = R_3 = R_4$

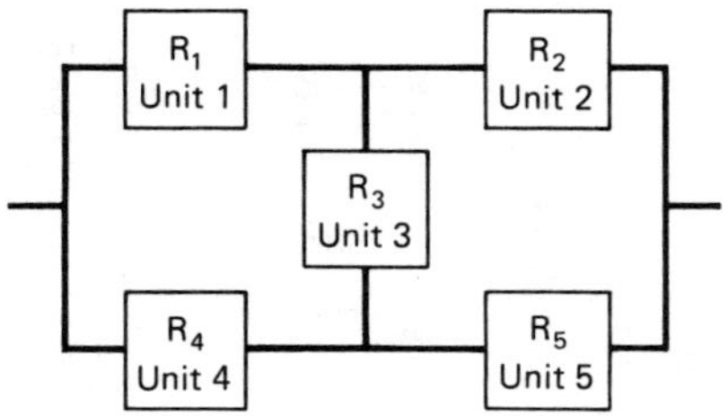	$R_n = 2R_1R_2R_3R_4R_5 - R_2R_3R_4R_5 - R_1R_3R_4R_5 - R_1R_2R_4R_5 - 2R_1R_2R_3R_5 - R_1R_2R_3R_4 + R_1R_3R_5 + R_2R_3R_4 + R_1R_4 + R_2 \cdot R_5$	$R_n = 2R^5 - 5R^4 + 2R^3 + 2R^2$ where $R = R_1 = R_2 = R_3 = R_4 = R_5$
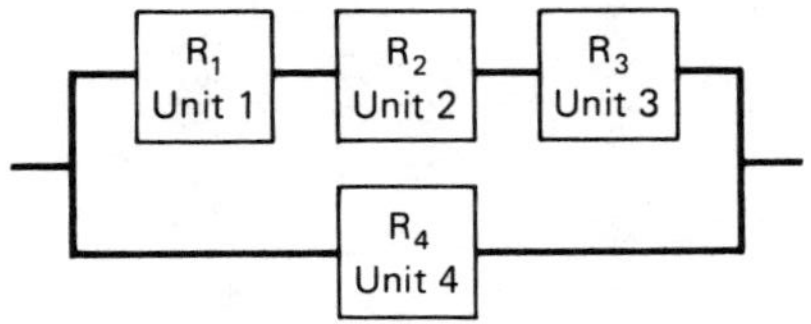	$R_n = 1 - (1 - R_4)(1 - R_1 \cdot R_2 \cdot R_3)$	$R_n = 1 - (1 - R)(1 - R^3)$ where $R = R_1 = R_2 = R_3 = R_4$
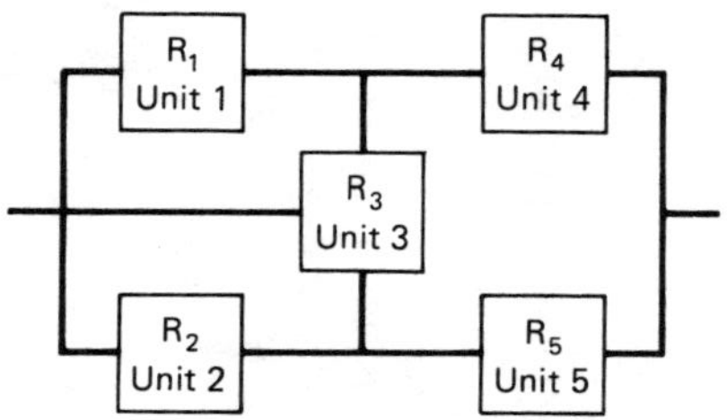	$R_n = [1 - (1 - R_4)(1 - R_5)]R_3 + (1 - R_3)[1 - (1 - R_1R_4)(1 - R_2R_5)]$	$R_n = [1 - (1 - R)^2]R + (1 - R)[1 - (1 - R^2)^2]$ where $R = R_1 = R_2 = R_3 = R_4 = R_5$

Table 3.5. Selective reliability functions and parameters.

Probability Distribution	Failure Density Function, $f(t)$	Reliability Function $R(t)$	Mean Time To Failure, MTTF	Hazard Rate, $\lambda(t)$
Exponential	$f(t) = \lambda e^{-\lambda t}$ for $t \geq 0$ where λ is the parameter, t is time.	$R(t) = e^{-\lambda t}$ where λ is known as the constant failure rate.	$\frac{1}{\lambda}$	λ
Rayleigh	$f(t) = \sigma t\, e^{-(\sigma t^2/2)}$ for $t \geq 0$ where σ is the parameter.	$R(t) = e^{-(\sigma t^2/2)}$	$\frac{(\pi)^{1/2}}{2\sigma}$	σt
Weibull	$f(t) = \frac{\sigma t^{\sigma-1}}{\beta^\sigma} \exp\left[-\left(\frac{t}{\beta}\right)^\sigma\right]$ for $t \geq 0$, $\beta > 0$, $\sigma > 0$. where σ is the shape parameter, β is the scale parameter.	$R(t) = \exp\left[-\left(\frac{t}{\beta}\right)^\sigma\right]$	$\beta\Gamma\left(\frac{1+\sigma}{\sigma}\right)$	$\frac{\sigma t^{\sigma-1}}{\beta^\sigma}$

where

λ is the parameter representing the constant failure rate of the component

t is time.

Obtain an expression for the component reliability.

Substituting equation (3.37) into equation (3.34) and integrating yields

$$R(t) = \int_t^\infty f(\tau)\, d\tau = \int_t^\infty \lambda\, e^{-\lambda\tau}\, d\tau$$

$$= \left[\frac{-\lambda e^{-\lambda\tau}}{\lambda}\right]_t^\infty \tag{3.38}$$

$$= e^{-\lambda t}$$

3.5.2. Mean Time To Failure

The mean time to failure (MTTF) of an item is given by

$$\text{MTTF} = \int_0^\infty R(t) \cdot dt \tag{3.39}$$

The derivation of the above relationship is shown in reference 11.

Table 3.5 presents the MTTF expressions associated with exponential, Rayleigh, and Weibull distributions. These MTTF expressions were developed with the aid of equation (3.39)

EXAMPLE 3.7

A mechanical component reliability is described by the following function:

$$R(t) = e^{-(1/\beta)t^2} \tag{3.40}$$

where

β is the scale parameter
t is time.

Obtain an expression for the component mean time to failure.

Substituting equation (3.40) into equation (3.39) and integrating, we get

$$\begin{aligned} \text{MTTF} &= \int_0^\infty e^{-(1/\beta)t^2}\, dt \\ &= \frac{\Gamma(1/2)}{2\left(\frac{1}{\beta}\right)^{1/2}} \qquad (3.41) \\ &= \frac{\sqrt{\pi}}{2} \cdot \left(\frac{1}{\beta}\right)^{-1/2} \end{aligned}$$

EXAMPLE 3.8

For $\beta = 4$, evaluate equation (3.41). From equation (3.41) we get

$$\begin{aligned} \text{MTTF} &= \frac{\sqrt{\pi}}{2}\left(\frac{1}{4}\right)^{-1/2} \\ &= 1.7725 \end{aligned}$$

3.5.3. Hazard Rate

The hazard rate, $\lambda(t)$, of an item is defined as follows:

$$\lambda(t) = \frac{f(t)}{R(t)} \tag{3.42}$$

where

$f(t)$ is the item failure density function
$R(t)$ is the item reliability function.

Table 3.5 presents the hazard rate expressions associated with three frequently used statistical distributions in reliability work known as exponential, Rayleigh, and Weibull.

EXAMPLE 3.9

The failure density function, $f(t)$, of an item is represented by

$$f(t) = \frac{\sigma t^{\sigma - 1}}{\beta^{\sigma}} \exp\left[-\left(\frac{t}{\beta}\right)^{\sigma}\right] \quad \text{for } \beta > 0, \quad \sigma > 0, \quad \text{and} \quad t \geq 0 \tag{3.43}$$

where

β is the scale parameter
σ is the shape parameter
t is time.

The above function is known as the Weibull failure density function. Obtain an expression for the Weibull hazard rate.

Using relationship (3.43) in equation (3.34), we get:

$$\begin{aligned} R(t) &= \int_t^{\infty} \frac{\sigma \tau^{\sigma - 1}}{\beta^{\sigma}} \exp\left[-\left(\frac{\tau}{\beta}\right)^{\sigma}\right] d\tau \\ &= \exp\left[-\left(\frac{t}{\beta}\right)^{\sigma}\right] \end{aligned} \tag{3.44}$$

Substituting relationships (3.43)–(3.44) into equation (3.42) yields

$$\begin{aligned} \lambda(t) &= \frac{\sigma t^{\sigma - 1}}{\beta^{\sigma}} \exp[-(t/\beta)^{\sigma}] \Big/ \exp\left[-\left(\frac{t}{\beta}\right)^{\sigma}\right] \\ &= \frac{\sigma t^{\sigma - 1}}{\beta^{\sigma}} \end{aligned} \tag{3.45}$$

3.5.4. Maintainability Function

When the duration of repair times is exponentially distributed, the maintainability function, $M(t)$, is given by

$$M(t) = 1 - \exp\left(-\frac{t}{\text{MTTR}}\right) \tag{3.46}$$

where

MTTR is the mean time to repair
t is the variable repair time
$M(t)$ denotes the probability that when repair begins at time $t = 0$, it will be accomplished successfully in time t.

For the exponential case, the mean time to repair is given by

$$\text{MTTR} = \frac{1}{\mu} \tag{3.47}$$

where

μ is the repair rate.

The mean time to repair is the mean of the distribution of equipment repair time and is estimated from the following equation:

$$\text{MTTR} = \sum_{j=1}^{m} \lambda_j T_j \Big/ \sum_{j=1}^{m} \lambda_i \tag{3.48}$$

where

T_j is the time needed to repair the equipment when the jth part fails
m is the number of repairable parts in the equipment
λ_j is the constant failure rate of the jth repairable part of the equipment.

EXAMPLE 3.10

A piece of equipment is composed of two replaceable subsystems. The subsystem I failure rate and replacement time are 0.005 failures/hour and 2 hours. Similarly, the subsystem II failure rate and replacement time are 0.01 failures/hour and one hour. Estimate the equipment mean time to repair.

The specified data the equation (3.48) simplifies to

$$\text{MTTR} = \frac{\lambda_1 T_1 + \lambda_2 T_2}{\lambda_1 + \lambda_2} \tag{3.49}$$

where

λ_1 is the subsystem I failure rate
λ_2 is the subsystem II failure rate
T_1 is the subsystem I replacement time
T_2 is the subsystem II replacement time.

Substituting the specified data into equation (3.49) results in

$$\text{MTTR} = \frac{(0.005)\ (2) + (0.01)\ (1)}{(0.005) + (0.01)} = \frac{0.01 + 0.01}{0.015} = 1.333 \text{ hours}$$

3.5.5. Steady State Availability and Unavailability

In reference 12, a system steady state availability, A_{ss}, is given by

$$A_{ss} = \frac{\mu}{\lambda + \mu} \tag{3.50}$$

where

λ is the system constant failure rate
μ is the system constant repair rate.

Similarly, the system steady state unavailability is given by

$$UA_{ss} = \frac{\lambda}{\lambda + \mu} \tag{3.51}$$

where

UA_{ss} is the system steady state unavailability.[12]

It should be noted that the sum of system steady state availability and unavailability is always equal to unity, i.e.

$$A_{ss} + UA_{ss} = \frac{\mu}{\lambda + \mu} + \frac{\lambda}{\lambda + \mu} = 1 \tag{3.52}$$

Since MTTF $= \frac{1}{\lambda}$ and MTTR $= \frac{1}{\mu}$, equations (3.50) and (3.51) yield

$$A_{ss} = \frac{\text{MTTF}}{\text{MTTR} + \text{MTTF}} \tag{3.53}$$

and

$$UA_{ss} = \frac{\text{MTTR}}{\text{MTTR} + \text{MTTF}} \tag{3.54}$$

EXAMPLE 3.11

Electronic system constant failure and repair rates are 0.003 failures/hour and 0.007 repairs/hour. Calculate the system steady state availability and unavailability.

Plugging the above specified data into equations (3.50) and (3.51) we get

$$A_{ss} = \frac{0.007}{0.003 + 0.007} = 0.7$$

and

$$UA_{ss} = \frac{0.003}{0.003 + 0.007} = 0.3$$

The system steady state availability and unavailability are 0.7 and 0.3, respectively.

3.6. RELIABILITY NETWORKS IN TIME DEPENDENT SITUATIONS

This section is concerned with reliability evaluation of networks in time-dependent situations, where it is assumed that the reliability of each network component depends on time. The only difference between the analyses of Section 3.4 of this chapter and the analyses of this section is that the Section 3.4 analyses assume that each network component reliability is constant rather than time-dependent (as in this section).

In this section, it is assumed that each network component reliability, from equation (3.33) is defined by

$$R(t) = \exp\left[-\int_0^t \lambda(\tau)\, d\tau\right] \tag{3.55}$$

If each network component failure rate is constant (i.e. $\lambda(t) = \lambda$), the component reliability from equation (3.36) is

$$R(t) = e^{-\lambda t} \tag{3.56}$$

EXAMPLE 3.12

A television set constant failure rate is 0.009 failures/hour. If the television set is switched on at time $t = 0$, calculate its reliability for 10 hours of operation.

Substituting the specified data into equation (3.56), we get

$$\begin{aligned} R(10) &= e^{-(0.009)(10)} \\ &= 0.9139 \end{aligned}$$

The reliability of the television set is 91.39%.

3.6.1. Series Network

With the aid of equation (3.6), the time-dependent reliability, $R_{ss}(t)$, of the series network is given by

$$R_{ss}(t) = R_1(t) \cdot R_2(t) \cdot R_3(t) \cdots R_n(t) \tag{3.57}$$

where

$R_i(t)$ is the ith unit time-dependent reliability for $i = 1, 2, 3, \cdots, n$.

If the failure rate of each series unit is constant, then substituting equation (3.56), into equation (3.57) results in

$$R_{ss}(t) = e^{-\lambda_1 t} \cdot e^{-\lambda_2 t} \cdot e^{-\lambda_3 t} \cdots e^{-\lambda_n t} = \exp\left[-\sum_{i=1}^{n} \lambda_i t\right] \tag{3.58}$$

where

λ_i is the ith unit constant failure rate for $i = 1, 2, 3, \cdots, n$
n is the total number of units in the series system.

Substituting equation (3.58) into equation (3.39) yields

$$\text{MTTF} = \int_0^\infty R_{ss}(t) \cdot dt = \int_0^\infty \exp\left[-\sum_{i=1}^{n} \lambda_i t\right] dt = \frac{1}{\sum_{i=1}^{n} \lambda_i} \tag{3.59}$$

EXAMPLE 3.13

An aircraft is composed of two independent and identical engines. Both engines must operate successfully for the aircraft to fly successfully. The constant failure rate of each engine is 0.004 failures/hour. Calculate the reliability of the aircraft for a 12 hour flying mission.

With the aid of specified data, from equation (3.58) we get

$$R_{ss}(12) = \exp\left[-\left(\sum_{i=1}^{2} \lambda_i\right)(12)\right] = e^{-(0.004+0.004)(12)} = 0.09085$$

Thus, the reliability of the aircraft flying successfully for the specified mission time is 90.85%.

3.6.2. Parallel Network

With the aid of equation (3.14), the time-dependent reliability, $R_p(t)$, of the parallel network is given by

$$R_p(t) = 1 - (1 - e^{-\lambda_1 t}) \cdot (1 - e^{-\lambda_2 t}) \cdots (1 - e^{-\lambda_m t}) \tag{3.60}$$

where

λ_i is the ith unit constant failure rate for, $i = 1, 2, 3, \cdots, m$
m is the total number of units in the parallel system.

For identical m units, equation (3.60) simplifies to

$$R_p(t) = 1 - (1 - e^{-\lambda t})^m \tag{3.61}$$

where

λ is each unit constant failure rate.

Substituting equation (3.61) into equation (3.39) results in

$$\begin{aligned} \text{MTTF} &= \int_0^\infty R_p(t) \cdot dt \\ &= \int_0^\infty [1 - (1 - e^{-\lambda t})^m]\, dt \\ &= \frac{1}{\lambda} \sum_{i=1}^{m} \frac{1}{i} \end{aligned} \tag{3.62}$$

EXAMPLE 3.14

An aircraft consists of three independent, active, and identical engines. At least one engine must operate successfully for the aircraft to fly successfully. The constant failure rate of each engine is 0.008 failures/hour. Compute the reliability of the aircraft for a 10 hour flying mission.

Substituting the given data into equation (3.61) yields

$$\begin{aligned} R_p(10) &= 1 - [1 - e^{-(0.008)(10)}]^3 \\ &= 1 - [1 - 0.9231]^3 \\ &= 0.9995 \end{aligned}$$

The reliability of the aircraft for the specified mission time is 99.95%.

3.6.3. Standby System

In the standby system one unit is active and k units are on cold standby. As soon as the operating (i.e. active) unit fails it is immediately replaced by one of the k cold standby (i.e. inactive) units.

In the standby system, it is assumed that (1) all the system units are independent and identical; (2) switching arrangement is perfect; (3) standby units remain as good as new in their standby mode; (4) failed units are never repaired; (5) each unit failure rate is constant. Thus, from reference 13 the reliability, R_s, of the $(k + 1)$ unit standby system is

$$R_s(t) = e^{-\lambda t}\left[1 + \lambda t + \frac{(\lambda t)^2}{2!} + \frac{(\lambda t)^3}{3!} + \cdots + \frac{(\lambda t)^k}{k!}\right] \tag{3.61}$$

where

λ is the each unit constant failure rate
t is time
k is the number of standby units.

Substituting equation (3.61) into equation (3.39) yields

$$\begin{aligned} \text{MTTF} &= \int_0^\infty R_s(t) \cdot dt \\ &= \int_0^\infty e^{-\lambda t}\left[1 + \lambda t + \frac{(\lambda t)^2}{2!} + \frac{(\lambda t)^3}{3!} + \cdots + \frac{(\lambda t)^k}{k!}\right] dt \qquad (3.62) \\ &= (k + 1)/\lambda \end{aligned}$$

EXAMPLE 3.15

A standby system is composed of two independent and identical units (i.e. one is operating, while the other is on standby). As soon as the operating unit fails, it is immediately replaced by the standby unit. It is assumed that the switching arrangement is perfect, the standby unit remains as good as new in its standby mode, and the failed unit is never repaired. If each unit constant failure rate is 0.007 failures/hour, compute the mean time to failure of the standby system.

Substituting the given data into equation (3.62), we get

$$\begin{aligned} \text{MTTF} &= \frac{k + 1}{\lambda} \\ &= \frac{1 + 1}{0.007} \\ &= 285.71 \text{ hours} \end{aligned}$$

Thus, the mean time to failure of the standby system is 285.71 hours.

3.6.4. Partial Active Redundancy

With the aid of equation (3.25), the time t, dependent reliability, $R_{rm}(t)$, of the r-out-of-m unit network is given by

$$R_{rm}(t) = \sum_{i=r}^{m} \binom{m}{r} (e^{-\lambda t})^i (1 - e^{-\lambda t})^{m-i} \tag{3.63}$$

where

λ is each unit constant failure rate.

Substituting equation (3.63) into equation (3.39) yields

$$\begin{aligned} \text{MTTF} &= \int_0^\infty R_{rm}(t)\, dt \\ &= \int_0^\infty \left[\sum_{i=r}^{m} \binom{m}{r} (e^{-\lambda t})^i (1 - e^{-\lambda t})^{m-i} \right] dt. \qquad (3.64) \\ &= \frac{1}{\lambda} \sum_{i=r}^{m} \frac{1}{i} \end{aligned}$$

2-out-of-3 Unit Network

For $r = 2$ and $m = 3$, from equation (3.63) we get

$$R_{rm}(t) = 3e^{-2\lambda t} - 2e^{-3\lambda t} \tag{3.65}$$

To compare the reliability of the 2-out-of-3 unit network with that of the single unit network with constant failure rate, we must find the crossover point. The crossover point is found by equating each network's reliability. Therefore, from equation (3.32) we get

$$R = \tfrac{1}{2} \tag{3.66}$$

which means that when the reliability of each unit is 0.5, the reliability of the 2-out-of-3 unit system is same as that of the single unit system.

Thus, to determine the constant failure rate of a unit, we plug R, from equation (3.66), into equation (3.56) as follows:

$$e^{-\lambda t} = \tfrac{1}{2}$$

Therefore

$$\lambda t = 0.693$$

which means that the 2-out-of-3 unit system will be more useful than a single unit system, only if the required normalized mission time, λt, is less than 0.693. If, however, it is greater than 0.693 it is better to have a single unit system rather than a 2-out-of-3 unit system. In Table 3.6 the reliability expressions for special cases of the r-out-of-m unit system are presented with the aid of equation (3.63).[14]

Table 3.6. Reliability expressions for *r*-out of-*m* unit configurations.

System	System Reliability
At least 1 out of 3 units must be working for success.	$R_s(t) = 3e^{-\lambda t} - 3e^{-2\lambda t} + e^{-3\lambda t}$ Where $R_s(t)$ is the system reliability λ is each unit constant failure rate t is time.
At least 3 out of 4 units must be working for success.	$R_s(t) = 4e^{-3\lambda t} - 3e^{-4\lambda t}$
At least 3 units out of 5 units must be working for success.	$R_s(t) = 6e^{-5\lambda t} - 15e^{-4\lambda t} + 10e^{-3\lambda t}$
At least $m - 1$ units out of m units must be working for success.	$R_s(t) = me^{-(m-1)\lambda t} - (m - 1)e^{-m\lambda t}$
At least 2 out of 4 units must be working for success	$R_s(t) = 3e^{-4\lambda t} - 8e^{-3\lambda t} + 6e^{-2\lambda t}$
At least $m - 2$ units out of m units must be working for success.	$R_s(t) = \binom{m}{2} e^{-(m-2)\lambda t} + (2m - m^2)e^{-(m-1)\lambda t} + \binom{m-1}{2} e^{-m\lambda t}$

EXAMPLE 3.16

An aircraft is composed of three independent, active, and identical engines. Each engine constant failure rate is 0.006 failures/hour. At least two engines must operate successfully for the aircraft to fly successfully. Calculate the aircraft reliability for nine flying hours.

Plugging the given data into equation (3.65), we get

$$\begin{aligned} R_{rm}(9) &= 3e^{-2(0.006)(9)} - 2e^{-3(0.006)(9)} \\ &= 2.6929 - 1.7009 \\ &= 0.9920 \end{aligned}$$

Thus, the aircraft reliability for the specified flying time is 99.20%.

3.7. RELIABILITY ESTIMATION

In order to extract useful information from the failure data of a piece of equipment, the data must be properly analyzed. Therefore, this section briefly introduces a graphical data analysis technique known as the *hazard plotting method.* This method can be used to find out which statistical distribution best fit to the failure data. In addition, the parameters of the statistical distribution involved can be estimated graphically.[15] Here the method is demonstrated for the Weibull distribution. From equation (3.45), the Weibull hazard rate is given by

$$\lambda(t) = \frac{\sigma t^{\sigma-1}}{\beta^{\sigma}} \tag{3.67}$$

According to reference 15, the cumulative hazard, $\lambda_c(t)$, of a distribution is given by

$$\lambda_c(t) = \int_{-\infty}^{t} \lambda(\tau)\, d\tau \tag{3.68}$$

Plugging equation (3.67) into equation (3.68) and integrating yields

$$\begin{aligned} \lambda_c(t) &= \int_0^t \frac{\sigma \tau^{\sigma-1}}{\beta^{\sigma}}\, d\tau \\ \lambda_c(t) &= t^{\sigma}/\beta^{\sigma} \end{aligned} \tag{3.69}$$

Taking the natural logarithms of equation (3.69) results in

$$\log_e \lambda_c = \sigma[\log_e t - \log_e \beta] \tag{3.70}$$

By rearranging equation (3.70) we get

$$\log_e t = \frac{1}{\sigma} \log_e \lambda_c + \log_e \beta \tag{3.71}$$

The left hand side of equation (3.71) is the linear function of $\log_e \lambda_c$. Therefore, from the plot of straight line equation (3.71), the values of parameters β and σ can be estimated. The slope of the straight line is equal to $1/\sigma$. Thus, the value of σ can be estimated from this fact. Similarly, at $\lambda_c = 1$ the value of β can be estimated from the fact that $\beta = t$. At this point, the value of the scale parameter β is equal to time t.

Similarly, one can estimate the values of the parameters of other statistical distributions occurring in reliability work.

3.8. SUMMARY

This chapter is basically concerned with mathematical aspects of reliability and maintainability. Reliability and maintainability histories are briefly discussed. Selective definitions and sources of current reliability and maintainability information are presented. The well known bathtub hazard rate curve is described, and each of the three distinct regions of the curve are discussed. The reliability networks such as series, parallel, r-out-of-m unit, and the standby system are presented. Reliability expressions for all these networks are developed. The reliability expressions for series and parallel configurations were developed with the aid of the decomposition technique. Formulas to obtain expressions for component reliability, mean time to failure (MTTF), and hazard rate are presented. The system maintainability function is defined. In addition, system steady state availability and unavailability formulas are given.

Mean time to failure expressions for series, parallel, standby system, and r-out-of-m unit networks are developed. The hazard plotting method to estimate values of parameters of the Weibull distribution is briefly described.

3.9. EXERCISES

1. Compare the developments of reliability and maintainability fields.
2. Define the following terms:

i. System mean time to failure.
ii. System unavailability.
iii. Partial active redundancy.
iv. System hazard rate.

3. What are the reasons for the useful life region of the bathtub hazard rate curve? What statistical distribution is applicable to represent the useful life region of that curve?
4. Develop an expression for reliability of the independent component network shown in Figure 3.15. Each component reliability is denoted by R. In Figure 3.15, the symbol R_i denotes the ith component reliability, for $i = 1, 2, 3, \cdots, 10$.

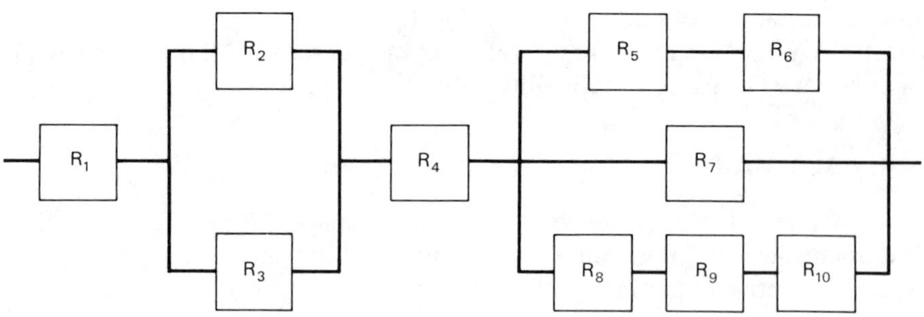

Figure 3.15. A reliability network.

5. Develop an expression for the reliability of the independent and identical component network shown in Figure 3.16. In the diagram the symbol R denotes the reliability of a component.

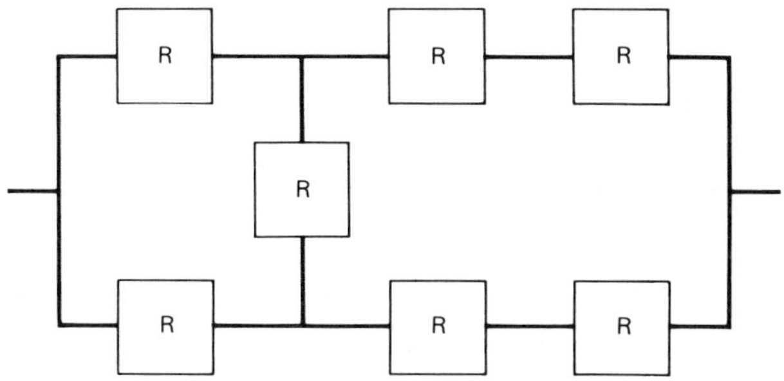

Figure 3.16. A complex network.

6. Prove that the mean time to failure of a redundant system is

$$\text{MTTF} = \frac{1}{\lambda} + \frac{1}{2\lambda} + \frac{1}{3\lambda} \tag{3.72}$$

where

λ is each unit constant failure rate.

7. An aircraft is composed of four independent, active, and identical engines. At least two of the engines must operate successfully for the aircraft to fly successfully. The constant failure rate of each engine is 0.009 failures/hour. Compute the reliability of the aircraft for a seven hour flying mission.
8. A standby system is composed of three independent and identical units (i.e. one is operating, while the other two are on standby). As soon as the operating unit fails, it is immediately replaced by one of the standby units. Similarly, when the second operating unit fails, it is immediately replaced by the last standby unit. It is assumed that the switching arrangement is perfect, the standby units remain as good as new in their standby modes, and the failed units are never repaired. Each unit constant failure rate is 0.001 failures/hour. Compute the standby system mean time to failure.
9. Prove that the reliability, $R_s(t)$, of a k identical, independent, and active unit system is given by

$$R_s(t) = ke^{-(k-1)\lambda t} - (k - 1)\, e^{-k\lambda t} \tag{3.73}$$

where

t is time
λ is each unit constant failure rate.

In this system at least $(k - 1)$ units out of k units must be working for system success.

10. An engineering component failure times are defined by the following failure density function:

$$f(t) = \frac{2}{\lambda} te^{-(1/\lambda)t^2}, \qquad \text{for} \qquad t \geq 0, \quad \lambda > 0 \tag{3.74}$$

where

λ is the scale parameter
t is time.

Develop an expression for the component hazard rate.

3.10 REFERENCES

1. Smith, S. A. "Service Reliability Measured by Probabilities of outage." *Electrical World,* Vol. 103, 1934, pp. 371–374.
2. Benner, P. E. "The Use of the Theory of Probability to Determine Spare Capacity." *General Electric Review,* vol. 37, 1934, pp. 345–348.
3. Lyman, W. J. "Fundamental Consideration in Preparing a Master System Plan." *Electrical World,* Vol. 101, 1933, pp. 778–792.
4. Smith, S. A. "Spare capacity Fixed by Probabilities of Outage." *Electrical World,* Vol. 103, 1934, 222–225.
5. Smith, A. A. "Probability Theory and Spare Equipment." *Edison Electric Inst. Bull.,* March 1934.
6. Dean, S. M. "Consideration Involved In Making System Investments For Improved Service Reliability." *Edison Electric Inst. Bull.,* vol. 6, 1938, pp. 491–496.
7. United States Department of Defense. MIL-STD-721B, *Definitions of Effectiveness Terms for Reliability, Maintainability, Human Factors and Safety.* Available from the Naval Publications and Forms Center, 5801 Tabor Avenue, Philadelphia, PA 19120.
8. Von Alven, W. H. *Reliability Engineering.* Prentice-Hall, Inc. Englewood Cliffs, New Jersey, 1964, pp. 5–19.
9. Kapur, K. C. "Reliability and Maintainability." In *Handbook of Industrial Engineering.* Edited by G. Salvendy, John Wiley and Sons, New York, 1982. pp. 8.5.1–8.5.34.
10. Thomason, T. *Introduction to Reliability and Quality.* Machinery Publishing Co. Ltd., London, 1969. pp. 50–51.
11. Kapur, K. C., Lamberson, L. R. *Reliability in Engineering Design.* John Wiley & Sons Ltd., New York, 1977, pp. 10–11.
12. Shooman, M. L. *Probabilistic Reliability: An Engineering Approach.* McGraw-Hill Book Company, New York, 1968, pp. 338–339.
13. Von-Alven, W. H. *Reliability Engineering,* Prentice-Hall, Inc., Englewood Cliffs, New Jersey, 1964, pp. 254–255.
14. United States Department of Defense. MIL-HDBK-217B, *Reliability Prediction of Electronic Equipment.* September 1974, pp. B-8.
15. Nelson, W. "Hazard Plotting for Incomplete Failure Data." *Journal of Quality Technology,* vol. 1, 1969.

CHAPTER 4

Introduction to Management

4.1 INTRODUCTION

Various definitions are used to describe management, and some examples are given in Figure 4.1. However, for our purpose, we will define management as an integrating process by which persons with authority create, maintain, and operate an organization in the selection and completion of its objective.[1]

The history of management goes back to the nineteenth century. In particular, most historians credit the last two decades of the nineteenth century for the real beginning of the management discipline. During this period, a recognized attempt to systematically investigate the development of basic management principles was started. In 1886, Henry R. Towne, president of the Yale and Towne Manufacturing Company, presented a paper entitled "The Engineer as Economist" at a meeting of the American Society of Mechanical Engineers. This is commonly regarded as the starting point of the search for a rational and systematic science of management. In his presentation Henry R. Towne emphasized the need of management as a field of independent study. However, Frederick W. Taylor is usually credited for initiating the scientific movement. In 1911, Taylor published a book entitled *The Principles of Scientific Management,* in which he questioned the traditional role of management.

Henry L. Gantt also played an important role in the development of management disciplines. His major contributions were the Gantt Chart and the task and bonus pay plan. In addition, he published the following three well-known books in 1910, 1916, and 1919, respectively:

1. Work, Wages, and Profits.
2. Industrial Leadership.
3. Organizing For Work.

During the same time frame, Harrington Emerson contributed to the management field in the areas of elimination of waste and creation of wealth. In 1912, Emerson helped establish the Efficiency Society of New York City.

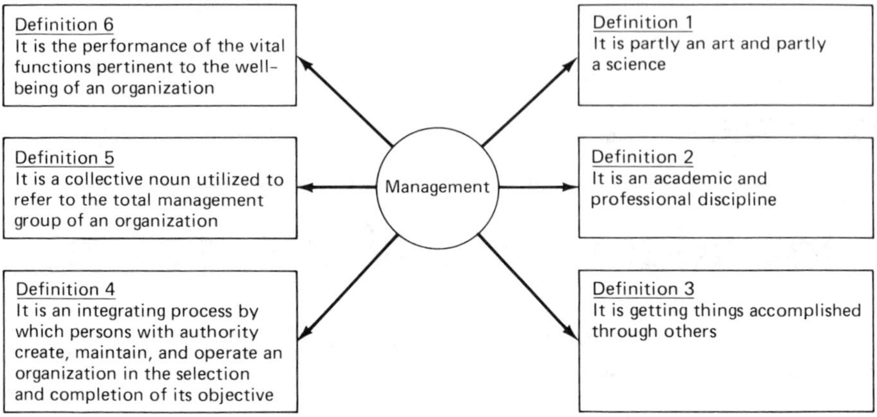

Figure 4.1. Examples of the management definition.

Frank B. and Lillian M. Gilbreth also made significant contributions to the development of the management field. The motions of the hand classifications into seventeen divisions are due to Frank Gilbreth. In addition, Frank Gilbreth published two books entitled *Motion Study,* and *Applied Motion Study* in 1911 and 1917, respectively. Lillian M. Gilbreth, noted for her contributions on the subject of human factors in industry, published a book in 1916 entitled *The Psychology of Management.*

A Frenchman named Henri Fayol also played an important role in the development of the management field. In 1916, he classified the study of management into various functional areas, namely, planning, controlling, organizing, and coordinating. His book entitled *Administration Industrielle et Generale,* was translated into English in 1930.

A very important event in the history of management took place in 1924 when a project known as the "Hawthorne study" was launched at the Western Electric Company plant in Illinois. This study was concerned with human factors and spanned a period of twelve years.

In the period between 1910–1925, societies were formed to aid the development of the management field. In 1914, the Society to promote the Science of Management was formed, and was the first professional management association. During the same decade, in 1916, the National Industrial Conference Board was established. Another significant step took place in 1923 when the American Management Association was formed.

Since the 1920's, various new developments have taken place in the management field. This chapter presents various aspects of the management discipline.

4.2 ORGANIZATIONS

There are various types of organizations, each with its own specific objective or objectives. An organization serves the following functions:[1]

1. It stores desirable information.
2. It allocates rewards to people (i.e., contributors and claimants).
3. It makes behavior of others more predictable.
4. It establishes a repetition of those actions which are desirable.
5. It makes certain that desired actions by members of organizations will be coordinated effectively.

According to reference 2, the following requirements influence the design of organizations:

1. Environments and organization philosophy.
2. Labor market and societal and technical systems' needs.
3. Needs, expectations, and goals of each member.
4. Needs of community, union-management, and social systems.
5. Requirements of social support systems, users, consumer, product, and market.
6. More than one objective concerning organization.

4.2.1 Organization Shape

Organization shape is influenced by factors such as the number of people supervised by each manager, and number of levels in an organization. For example, an organization with a large number of levels will have smaller spans of control (provided other things are equal). That is, an organization with a large number of levels, has long communication channels and undesirable administrative expenses. On the other hand, if an organization has too few organizational levels, the spans of control become too big and the effectiveness of supervision will be hampered.

The number of organizational levels also depends on the size of the organization. For example, a survey of 128 business organizations revealed that companies with 100, 1,000, and 10,000 employees had approximately four, six, and between seven and eight organizational levels, respectively.[1]

Another factor that influences the number of levels in an organization is technology, although, its influence is not as defined, as is the size influence on an organization. At least one study has demonstrated that as an organization becomes more technologically complex, its number of supervision levels increases.[1]

4.2.2 Organization Planning

Careful consideration is needed when planning an organization. According to reference,[3] the following points must be given careful consideration when planning an organization:

1. Making the line of authority and responsibility clear from the top to bottom of the organization.
2. A person delegating responsibility is accountable.
3. Structuring the organization so that the organizational levels are minimal.
4. The delegation of authority and responsibility is as near as possible to the point of action.
5. Each part of the organization accomplishes its own objective as effectively as its parent enterprise.
6. Positions are designed so that the functional and managing duties are separate.
7. Each member in the organization receives orders from one higher authority only.

4.2.3 Factors Influencing the Span of Supervision

According to the experience of Lockheed Missiles and Space Company, there are seven factors that influence the span of supervision.[4] These factors are shown in Figure 4.2.

4.2.4 Organizational Diagrams and Methods of Organization

This section briefly explains the organizational charts and methods of organization. Organizational diagrams are widely used by business organizations for the following reasons:

1. Promotes understanding of communication difficulties.
2. Depicts basic relationships and authority in an organization.
3. Helps to identify weak or indefinite control.
4. Orients newcomers to structural relationships and complexities.
5. Assigns responsibility.
6. Serves as a visibility tool by providing the full sweep of organizational activities and services.

There are various methods used to establish an organizational structure of a firm. Some of these methods are as follows:

1. Functional organization.
2. Goal-oriented organization.
3. Matrix organization.

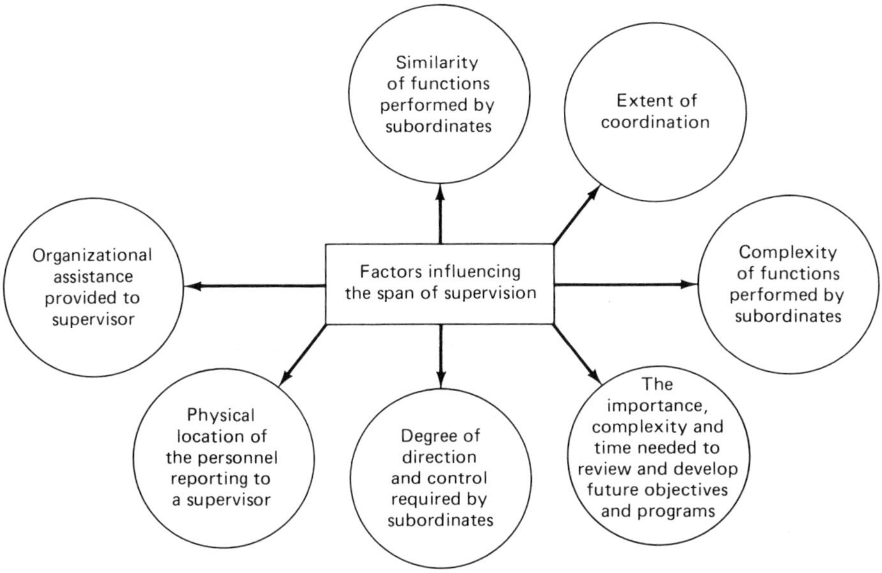

Figure 4.2. Factors influencing the span of supervision.

Functional Organization

The functional organization involves the pooling of many people who perform similar work in the same department. Figure 4.3. shows an example of organization by function.

According to reference 5, the organizations with mostly long-term jobs tend to favor the functional organization approach. In addition, large organizations and research groups also favor this approach.

There are various advantages and disadvantages of the organizational approach. Some of the advantages of the approach are as follows:

1. A good potential for efficiency.
2. Resources are utilized fully.
3. People can learn additional skills through interactions with others that are doing similar work.
4. Duplication of facilties is eliminated.
5. Even distribution of work load.
6. More uniform products and consistent policy.
7. Easier group supervision because of homogeneity.

Similarly some of the disadvantages of this method are as follows:

1. Slower work flow.
2. Multiple supervision.

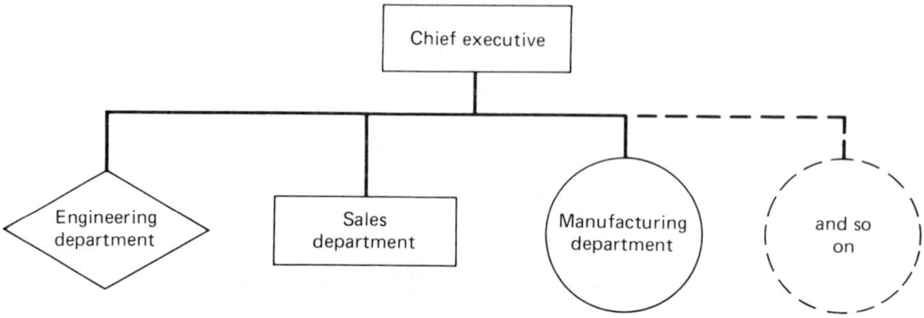

Figure 4.3. A functional organization.

3. Frequent conflict between departments.
4. Difficult scheduling and coordination between departments.

Some of the uses of the functional organization approach are as follows:[1]

1. Conditions where technology is routine and relatively fixed.
2. Situations where the market is stable and predictable.
3. Situations where size is big enough to allow standardized equipment and jobs.

Goal-Oriented Organization

The goal-oriented organization approach is quite distinct from the organization by function method, in that it is adaptive and creative but not particularly efficient. The goal-oriented unit consists of people with diverse specialties, and is headed by a leader who can integrate the efforts of people of different specialties.

This type of organizational structure is used to complete projects, to solve unique problems, and so on. Figure 4.4 shows an example of a goal-oriented structure. Some of the advantages of the goal-oriented organizations are as follows:

1. Tendancy to meet schedules effectively.
2. Provides a training ground for upper management of the organization.
3. Efficient flow of work.
4. Focuses attention on one project.
5. A vehicle for teamwork.
6. A tool for specialization by product.

Similarly some of the disadvantages of this approach are as follows:

1. Duplication of manpower and facilities.
2. Difficult to advance specialists within their field because of isolation from other specialists in the same field.
3. Less uniformity.

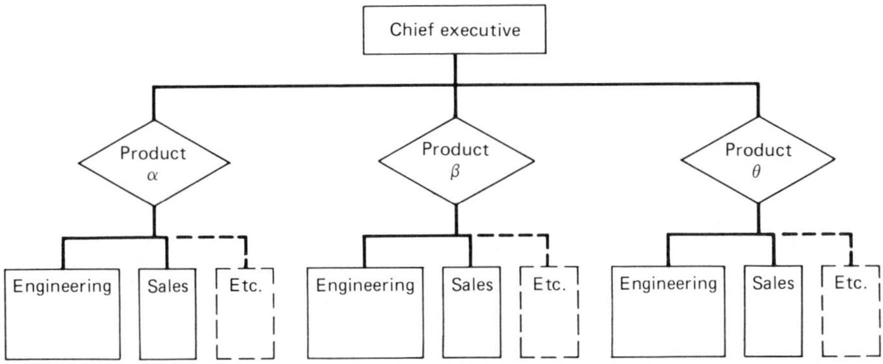

Figure 4.4. A goal-oriented structure.

Goal-oriented organization is useful in the following cases:

1. Creativity and adaptiveness are the basis for the organizational survival.
2. Unstable market.
3. Market is difficult to predict.
4. Technology is relatively made to order.

Matrix Organization

Matrix organization is a hybrid form of organizational structure that results from combining functional organization and goal-oriented organization. The matrix organization method has more advantages than the previous two methods and came about because some firms were not big enough to utilize, effectively, the project organizational approach. Figure 4.5 is an example of a matrix organizational structure. As exhibited by the matrix organization diagram, the project manpower is on loan to the project manager from various functional departments, and is responsible to both project and functional managers.

The matrix organizational approach negates some of the principles of management. Two such examples are as follows:

1. The program manager has more responsibility than authority, which is contrary to the management principle calling for equal responsibility and authority.
2. The people assigned to a goal-oriented unit have more than one boss. Again, this is contrary to the management principle calling for one boss for each person.

Some of the benefits of matrix organization are as follows:

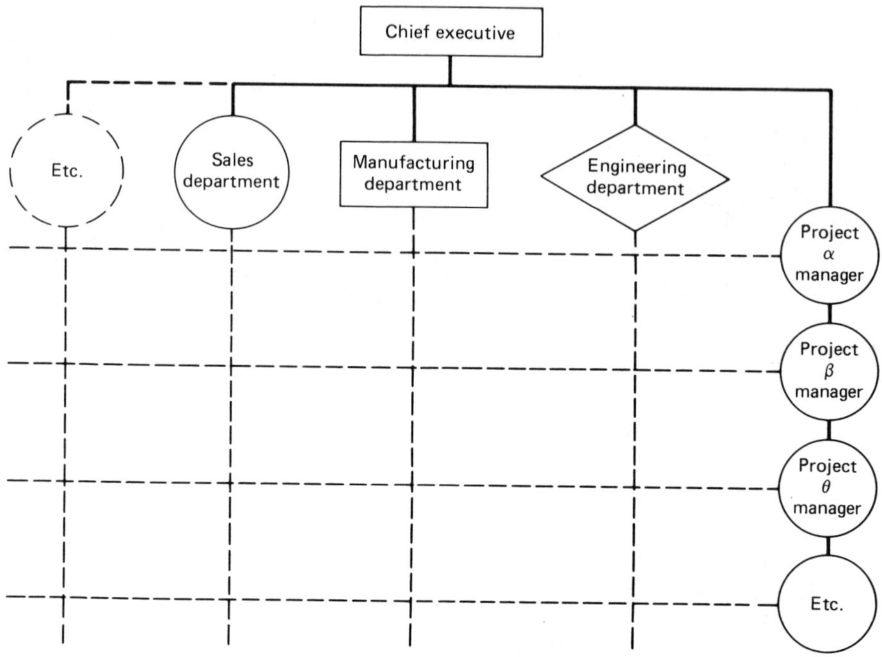

Figure 4.5. A matrix organization.

1. Enhances the role played by middle management.
2. It is an efficient approach for making decisions.
3. It provides more effective control.

One of the important drawbacks of this method is that the probability of having surplus skills increases in the event of a shortage of projects. However, the matrix organization is suitable in cases such as follows:

1. In situations where in comparison to maximum economy, the time constraints are more important.
2. In situations where temporary project groups are used. (The groups will be terminated at the completion of projects.)

4.3 DEPARTMENT FUNCTIONS OR RESPONSIBILITIES

This section briefly discusses the functions or responsibilities of reliability, maintainability, quality control, and maintenance engineering departments.

4.3.1. Responsibilities of a Reliability Department

The principal responsibilities of a reliability engineering department are as follows:

1. Reliability evaluation and allocation.
2. Reliability planning and providing input to design specifications and proposals.
3. Training reliability manpower.
4. Reliability research and development.
5. Reviewing design and specification with respect to reliability.
6. Monitoring subcontractor (if applicable) reliability activities.
7. Auditing the reliability activities.
8. Collecting and analyzing the failure data.
9. Reliability demonstration and reliability growth monitoring.
10. Consulting.

4.3.2. Functions Performed by the Maintainability Organization

Some of the functions performed by the maintainability organization are as follows:

1. Developing maintainability program plan, policies, and procedures.
2. Staffing the maintainability engineering effort.
3. Participating in design reviews.
4. Training maintainability manpower.
5. Maintainability allocation and prediction.
6. Preparing maintainability demonstration plans.
7. Performing analysis of the maintainability field data.
8. Developing maintainability design reports.
9. Providing consulting services to others.
10. Developing the maintainability data bank.
11. Performing maintainability trade-off studies.
12. Documenting maintainability related information.
13. Monitoring subcontractors (if applicable) with respect to maintainability activities.

4.3.3. Activities of a Quality Control Department

The responsibilities and activities of a quality control department may vary from one department to another. However, many of the quality control department activities are as follows:

1. Quality and inspection planning.
2. Process and measurement control.
3. Inspection.
4. Statistical analysis.
5. Performing analysis of quality field complaints, causes of defects, and quality costs.

6. Preparing programs to prevent defects.
7. Quality auditing and monitoring vendors.
8. Training quality manpower.
9. Performing research and development studies.

4.3.4. Responsibilities of a Maintenance Engineering Department

Some of the responsibilities of a maintenance engineering department are as follows:

1. Maintenance planning.
2. Execution of maintenance of existing equipment.
3. Management of maintenance manpower.
4. Equipment inspection.
5. Equipment installation.
6. Modification of existing equipment.
7. Generation and distribution of utilities.
8. Administration.
9. Keeping essential maintenance data for future studies.
10. Performing analytical studies.
11. Providing input to the design of new equipment.

4.4. GROUP BRAINSTORMING

To obtain new ideas, various creative techniques are used such as *group brainstorming.* This method, developed by Alex Osborn for modern day applications, is probably the most widely known creativity method.[8] As the title of the technique indicates, a group of people take part in a brainstorming session. During a session, one problem solving idea triggers another idea, and the process continues. The following points should be kept in mind during group brainstorming:

1. Strive for the quantity of ideas.
2. Bar criticism of any idea.
3. Ensure that each session is less than one hour.
4. Welcome "free wheeling" during the session.
5. Encourage modification or combination with another idea.
6. Ensure that the people taking part in the brainstorming session have different backgrounds but similar interests. (The rank of the people should be fairly equal.)
7. Ensure that ideas spelled out during the session are recorded.

In 1958, C. W. Taylor published the result of a study in which he compared individual creation with group brainstorming.[9] His results indicated the following:

1. A group of people generate more ideas than one individual.
2. The members of a group working individually, generate more ideas than when working as a group.
3. The members of a group working as individuals, generate ideas of higher mean quality.

4.5. MATHEMATICAL MODELS

Various types of mathematical models are being used in management work. In mathematical models, symbols are used to represent an entity rather than a physical device. In operations research studies, mathematical models are the most widely used models. According to reference 1, some of the properties of a good operations research model are as follows:

1. Generates solutions that can be used.
2. Is cost effective.
3. Provides non-trivial insights.
4. Is easy to comprehend.
5. Is efficient with respect to computing time.
6. Is concerned with problems that are useful to the manager.
7. Provides solutions within the time frame required.
8. Provides flexibility.

The classifications of mathematical models are shown in Figure 4.6. Brief descriptions of classifications shown in Figure 4.6 are given in reference 1. Various types of selective mathematical models are presented below.

4.5.1. Financial Ratios

This section presents various selective financial ratios useful to evaluate the financial health of a company. More clearly, these ratios are used to determine the financial strength and operating efficiency of an organization. The calculated values of these ratios can be compared to the earlier calculated values to detect trends, to the ratios from similar organizations, or to industry averages. Some of these ratios are as follows.

RATIO I

This is known as the acid test ratio, indicates a company's ability to pay its current debts without selling its inventory. The ratio is defined as follows:

$$R = \frac{A_c - I}{L_c} \tag{4.1}$$

where

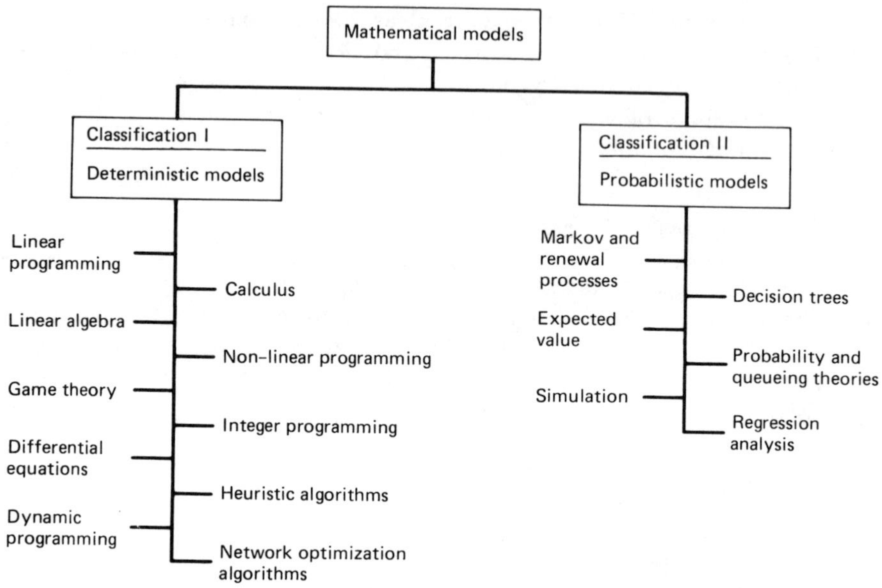

Figure 4.6. Classifications of mathematical models.

R is the acid test (quick) ratio
L_c is the current liabilities
I is the inventory
A_c is the current assets.

RATIO II

This ratio is known as the profitability ratio and is used to compute gross profit margin. Thus, the gross profit margin is given by

$$P_m = \frac{S - C_{gs}}{S_n} \tag{4.2}$$

where

P_m is the gross profit margin
S is the sales
S_n is the net sales
C_{gs} is the cost of goods sold.

RATIO III

This is known as the current ratio and is given by

$$R_c = \frac{A_c}{L_c} \tag{4.3}$$

where

R_c is the current ratio
A_c is the current assets
L_c is the current liabilities.

4.5.2. New Product Development Indices

Each day newly developed products are put into market and thus, manage ment is concerned with selecting the right new product idea out of many. According to reference 10, 50–500 new product ideas are required at the initial exploration stage, for a new product to be introduced successfully. There are various indices that management uses to determine the worth of a proposed new product with respect to other proposed products. Two of such indices are as follows.

INDEX I

This is known as the project value index and is given by

$$I_{pv} = T_s \cdot C_s \cdot V_a \cdot D_p(L_u)^{1/2}/C_{tp} \tag{4.4}$$

where

C_{tp} is the total cost of the project
L_u is the useful life of the product in years
I_{pv} is the project value index
V_a is the product sale in units
D_p is the profit per unit
C_s is the product's chances for commercial success (a rating scale from 0–10)
T_s is the product's chances for technical success (a rating scale from 0–10).[11]

INDEX II

This is known as the *index of relative worth* and is given by[11]

$$I_{rw} = Q\,R/k \tag{4.5}$$

where

I_{rw} is the index of relative worth
k is the cost of future research
R is the net return for an arbitrary period of five years
Q is the probability of commercial achievement of the objective.[11]

4.5.3. Break-Even Analysis

This is a widely used technique to study the relationship between total incomes and total costs at potential levels of operation. The break-even point can be found either graphically or analytically. The formula to find the break-even point is as follows

$$N = \frac{C_f}{P_s - C_v} \tag{4.6}$$

where

C_f is the fixed cost
C_v is the variable cost/unit
P_s is the sale price/unit
N is the break-even point (i.e. number of units).

EXAMPLE 4.1

The cost of manufacturing an engineering component is \$50,000 (i.e., fixed cost). In addition, the variable cost is \$6/component, and the selling price of each component is \$8. Calculate the number of units to be sold to break-even.

By substituting the specified data into equation (4.6), we get

$$N = \frac{50{,}000}{8 - 6} = 25{,}000 \text{ units.}$$

This means that 25,000 units must be sold to break-even.

4.6. SUMMARY

This chapter briefly describes various aspects of general management. The history of management is briefly discussed and various definitions of management are presented. The next topic discussed in the chapter is concerned with organizations; namely the factors influencing the design of organizations, organization planning, span of supervision, and organization diagrams. In addition, the three methods of organization, functional organization, goal-oriented organization, and matrix organization, are discussed.

Functions or responsibilities of departments such as reliability engineering, quality control, and maintenance engineering are outlined. The method of group brainstorming is discussed briefly.

The last topic presented in the chapter is concerned with mathematical models. Included in this chapter are two classifications of mathematical models, a number of properties of a good operations research model, and three well-known financial ratios (quick, current, and profitability ratios). In addition, two new product development indices known as the project value index and relative worth index are defined. Finally, the break-even analysis technique is explained, briefly.

4.7. EXERCISES

1. Write down a most commonly used definition of management and compare it with at least three other definitions of management.
2. What are the factors that influence the shape of an organizational structure?
3. Write down a number of useful guidelines for organization planning.
4. Discuss the factors influencing the span of supervision.
5. Compare the following two types of organizations:
   i. Matrix.
   ii. Goal-oriented.
6. What are the functions of a safety engineering department?
7. What are the main differences between the activities of the maintenance and maintainability engineering departments?
8. List the rules to be observed during a brainstorming session.
9. What are the characteristics of a good operations research model?
10. An engineering company wishes to manufacture an engineering item. The fixed and variable costs will be \$70,000 and \$15/item, respectively. The estimated selling price of each item is \$20. Calculate the number of items the company must sell to break-even.

4.8. REFERENCES

1. Bittel, L. R. *Encyclopedia of Professional Management.* McGraw-Hill Book Company, New York, 1978.
2. Davis, L. E. "Organization Design." In *Handbook of Industrial Engineering,* edited by G. Salvendy, John Wiley & Sons, New York, pp. 2.1.1–2.1.2.
3. Karger, D. W., Murdick, R. G. *Managing: Engineering and Research.* Industrial Press Inc., 200 Madison Ave., New York, 1969.
4. Stieglitz, J. "Optimizing Span of Control." *Management Record,* Vol. 24, 1962, pp. 25–29.
5. Tangerman, E. J. "Engineering Organization Charts of 25 Modern Companies." In *Management Guide for Engineers and Technical Administrators,* edited by N. P. Chironis, McGraw-Hill Book Company, New York, 1969, pp. 24–25.
6. Headquarters, United States Army Material Command. *Maintainability Engineering Theory and Practice,* AMCP 706–133, 5001 Eisenhower Ave., Alexandria, VA 22333, USA., / January 1976, pp. 3.1–3.3.
7. Juran, J. M., Gryna, F. M. *Quality Planning and Analysis.* McGraw-Hill Book Company, New York, pp. 562.

8. Osborn, A. *Applied Imagination.* Scribner, New York, 1957.
9. Taylor, C. W. "Some Variables Functioning in Productivity and Creativity." In Proceedings of Research Conference on the Identification of Creative Scientific Talent, edited by C. W. Taylor. University of Utah Press, Salt Lake City, Utah, 1958.
10. Uman, D. B. *New Product Programs.* American Management Association, Inc., New York, 1969.
11. Heyel, C. *The Encyclopedia of Management.* Van Nostrand Reinhold Company, New York, pp. 613.

CHAPTER 5

Engineering Manpower

5.1. INTRODUCTION

The various engineering professionals who directly or indirectly contribute to a product's reliability and maintainability are design engineers, reliability engineers, quality control engineers, safety engineers, human factors engineers, and so on. These professionals are required to have certain characteristics, qualities, attributes, or qualifications to perform their function effectively, and to ensure that their resulting output is at the desired level or quality. Should these professionals be lacking in the above mentioned characteristics, the product's reliability and maintainability may be effected, and the product under development may not be delivered according to plans. Factors such as reliability, maintainability, and delivery date are important because they influence the life cycle cost of the product. This chapter presents the qualities, qualifications, attributes, characteristics, and functions of the following selective professionals:

1. Engineer.
2. Engineering manager.
3. Project engineer.
4. System engineer.
5. System project manager.
6. Design engineer.
7. Contract negotiator.
8. Quality control manager.
9. Configuration manager.
10. Reliability engineer.
11. Design auditor.
12. Product safety director.
13. Maintainability engineer.
14. Human factors engineer.
15. Personnel officer.

5.2. CHARACTERISTICS AND NEEDS OF AN ENGINEER

An engineer plays an important role in the reliability and maintainability of equipment. His characteristics and needs may directly or indirectly influence the reliability and maintainability of the product for which he is responsible; therefore, this section examines the characteristics and needs of a typical engineer. Some of the characteristics of an engineer are as follows:

1. He is conservative and works hard.
2. He is honest, sincere, and family oriented.
3. He is well-motivated and organized.
4. He is reserved, independent, and has a small number of close friends.
5. He is cautious, tends to be conventional, and is basically interested in mechanically oriented hobbies.
6. He is a poor communicator and has high visual ability.
7. He is emotionless, friendly, and very intelligent.
8. He is basically a shy person and possesses little interest in people.

The needs of an engineer are another factor that may directly or indirectly affect the reliability and maintainability of a product for which he is responsible. Some of these needs are:[1]

1. Adequate work facilities, and stimulating and challenging work.
2. Work variety and correct work assignment.
3. Economic advancement and opportunities for self-development.
4. Independence to tackle a work problem and the opportunity for ideas to be put into practice.
5. Availability of necessary technical assistance and necessary supporting staff.
6. Supervision by competent bosses.
7. Employment with a reputable organization with clearly defined responsibility and authority.
8. Recognition of efforts from the management.

5.3. ACTIVITIES AND ATTRIBUTES OF AN ENGINEERING MANAGER

This section describes the activities and qualities of an engineering manager. An engineering manager must engage in various activities. Some of them are as follows:

1. Organizing and planning.
2. Staffing and communicating.
3. Personnel development.
4. Training and counselling.[2]

A successful engineering manager possesses certain qualities. Some of these qualities are as follows:

1. Ability to reason and flexibility
2. Tolerance and ability to listen
3. Self-confidence and empathy
4. Ability to control emotion
5. Quick to praise and criticize
6. Independent from suspicion and prejudice
7. Ability to recognize different point of views
8. Readiness to give credit to others
9. Quick to see good in others
10. Ability for self appraisal
11. Flexibility
12. Fairness[3]

5.4. BASIC QUALIFICATIONS OF A SYSTEM ENGINEER AND A SYSTEM PROJECT MANAGER

In order to produce effective engineering systems, both the system engineers and the system project managers must be developed to their full potential. This section discusses the basic qualifications of a system engineer and a system project manager. Some of the basic qualifications that a system engineer should have are as follows:[4]

1. Competence in methodology and procedures for formulating and finding solutions to system-level problems.
2. A satisfactory working knowledge of disciplines such as mathematics, physics, electronics, economics, physiology, and programming.
3. Ability to sort and sift facts.
4. Ability to conceive of and plan applications that combine the optimum manpower, equipment, approaches, and supporting techniques to fulfil the specified goal.
5. Thorough knowledge of preparing and managing budgets, cost estimates,and project management schedules.
6. Thorough knowledge of preparing and managing engineering design, development, and test programs.
7. Thorough knowledge of preparing and managing software design, development, and test programs.
8. Thorough knowledge of preparing and managing the engineering methods used to translate system needs into resulting item designs, and plans for selection and training of required system specialist manpower.

9. Thorough knowledge of preparing and managing plans for acquisition and manufacturing of system components, construction of required facilities and so on.

Similarly, some of the basic qualifications of a system project manager, in addition to being a successful system engineer, are as follows:

1. Rejects sloppy or poor performance.
2. Readily enters into discussions from which problems may surface.
3. Encourages the maintenance of technical integrity.
4. Possesses confidence in his subordinates and associates.
5. Confident to direct the system development effort.
6. Does not allow an inflexible approach to a specified design occur.
7. Always looks out for staff members who posssess superior knowledge in specific areas.

5.5. OBJECTIVES AND AUTHORITIES OF A PROJECT ENGINEER

The project engineer plays an instrumental role in the successful completion of an engineering project. He has the vested authority and responsibility to manage a project in such a way that the project objectives are accomplished successfully. Usually engineering projects involve a contract between two or more organizations; therefore the contractor's project engineer has the following basic contract objectives:[5]

1. To deliver an end product fully meets the specified requirements and the delivery schedule.
2. To satisfy the company profit objective effectively.

In addition, some of the areas that require decision-making authority are as follows:

1. Selection of sub-contractors.
2. Choosing manpower.
3. Expenditure of budgeted money.
4. Identification of types of tests.
5. Scheduling resources and manpower.
6. Manufacture or procure decision.
7. Selection of parts.
8. Directing the design method.

5.6. DESIGN MANPOWER

This section discusses two important people involved in engineering design, namely, the design engineer and the design auditor. The design engineer is the backbone of product design engineering. His basic concern is to produce a successful design at the lowest cost. Some of the characteristics of a design engineer are as follows:[6]

1. Has a broad intellectual scope and tolerates criticisms.
2. Has good speaking and writing skills.
3. Possesses a keen visual sense, a high degree of self-confidence, and a great deal of initiative and concentration.
4. Has no faith in routine work and takes specific calculated risks.
5. Very creative, observant, and curious.
6. Has a good grasp of basic principles.
7. Has a good memory, is flexible, and persistent.
8. Has the ability to integrate.

The design auditor performs the function of design auditing. He is a counterbalancing generalist in a big and growing population of specialists. Actually, the job of design auditor has arisen due to the increasing specializa tion of software, electronics, and systems designs. Independence and impartiality are important characteristics that the design auditor should possess. The steps shown in Fig. 5.1 are found to be most useful to perform design auditing.[7]

The following guidelines are considered useful to the design auditor:

1. Make certain that errors disclosed are handled judiciously.
2. Make sure that usually only the concrete recommendations are seen by other people.
3. Make certain that the auditing is conducted impartially.
4. Make sure that you are thoroughly discreet in your activities.
5. Make it clear to everyone involved that the auditor's objective is not to fix blame for honest mistakes.

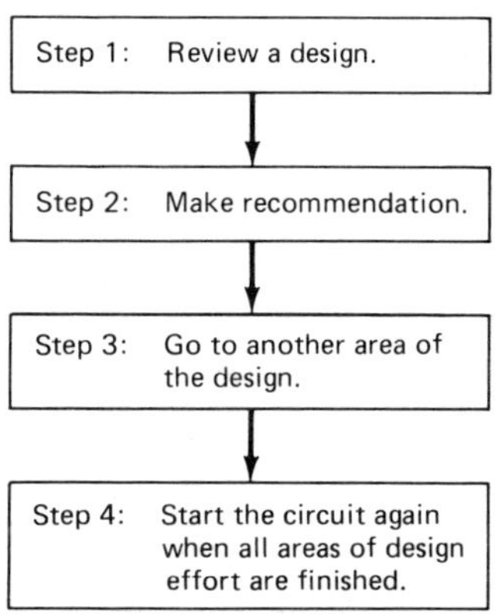

Figure 5.1. Steps for design auditing.

6. Perform auditing function early in the program so it does not delay the project.

According to reference 7, the following individual level assignments are accomplished when conducting reliability audits:

1. Developing the objective of the design audit.
2. Outlining the procedures for the design audit.
3. Organizing the effort associated with the design audit.
4. Conducting the design audit.
5. Reviewing the end results of the design audit.
6. Analyzing the audit performance.
7. Regulating the operation of the design audit.

5.7. CONFIGURATION MANAGER AND CONTRACT NEGOTIATOR

This section discusses two important professionals involved with equipment configuration and engineering contract negotiation, namely, the configuration manager and the contract negotiator. The configuration manager provides indirect input to improve equipment reliability and maintainability. Broadly speaking, the responsibilities assigned to a configuration manager may be divided into four categories:[8]

1. Operations of configuration control board.
2. Management of data.
3. Configuration planning.
4. Configuration administration.

In order to fulfil the above responsibilities effectively, the configuration manager should have the following qualities:

1. Familiarity with the company and its operations.
2. Ability to organize.
3. Administrative and documentation skills.
4. Attention to details.
5. Knowledge of quality control, design, manufacturing, and testing.
6. Familiarity with the equipment in question.
7. Tactfulness.
8. Diplomacy and aggressiveness.

The job of the contract negotiator is also very important, because his competence plays a critical role in winning a contract. Some of the qualities of a contract negotiator are as follows:[9]

1. Impersonality and patience.
2. Objectivity and fast thinking mind.
3. Effectively expresses thoughts and arguments.
4. Effectively translates financial and technical information into suitable language.
5. Takes the necessary measures after considering the current situation.

In addition a good negotiator:

1. Treats his counterparts as human beings.
2. Is well prepared for negotiation.
3. Is alert thoughout the negotiation of a contract.
4. Is prepared for changes.
5. Surveys the contractor's facilities.
6. Has the capacity to listen to other people.
7. Is prepared for a battle of wits.

5.8. QUALITY CONTROL MANAGER AND QUALITY CONTROL PERSONNEL

This section describes the qualities and functions of a quality control manager. In addition, responsibilities of quality control personnel are also presented.
Some of the functions of a quality control manager are as follows:[10]

1. Coordination and quality planning.
2. Supervision of analytical and inspection personnel.
3. Consulting.
4. Liaison with outside bodies.
5. "Reporter and scorekeeper" for the firm with respect to quality.

A good quality control manager has the following attributes:

1. Good knowledge of statistical analysis.
2. Effective in utilization of time.
3. Good planner, motivator, and listener.
4. Good knowledge of research, engineering, manufacturing, finance, and marketing.
5. Capable of organizing, delegating, and selling.
6. Creative, fair and honest.
7. Patience, enthusiasm, and the ability to communicate.
8. Good health and the ability to think effectively.

The responsibilities of quality control manpower may vary from one place to another due to factors such as the type of product, the degree of automation, the size of the plant, plant layout, and so on. However, some of the basic responsibilities of quality control personnel are as follows:[11]

1. Coordinate other departments' quality functions.
2. Evaluate the quality of vendor systems and procured parts.
3. Develop and apply necessary inspection plans.
4. Perform analysis of quality associated defects.
5. Develop quality standards.
6. Use statistical quality control techniques.
7. Establish receiving, outgoing, and in-process inspections, according to running quality.
8. Attend and conduct quality related meetings.
9. Audit the quality control system.
10. Carryout quality functions according to procedures and recommendations.
11. Avoid "firefighting" on quality related problems.

5.9. RELIABILITY AND MAINTAINABILITY ENGINEERS

This section briefly discusses the two most important professionals concerned with the reliability and maintainability of equipment, namely, the reliability and maintainability engineers. Broadly speaking, the reliability engineer assists management in defining, evaluating, and containing its risks, through the following steps:

1. Performing analysis of customer requirements.
2. Determining the accuracy of reliability predictions.
3. Evaluating the reliability of prediction plans.

Some of the functions of a reliability engineer are as follows:

1. Develop reliability growth monitoring procedures.
2. Provide information on parts' failure rates.
3. Predict the effect of environment on item reliability.
4. Evaluate the reliability of a product.
5. Providing information on parts' reliability requirements.
6. Develop reliability evaluation models and methods.
7. Develop reliability demonstration and evaluation test procedures.
8. Monitor subcontractors' reliability program.
9. Provide consulting services to others.
10. Participate in design reviews.

A reliability engineer, like any other professional engineer, is a well qualified engineer. In some cases, reliability engineers are former design engineers, who gain knowledge of reliability engineering through special courses or by other means.

The maintainability engineer is also a well qualified engineer, whose functions may vary from one company to another depending on the type of equipment being manufactured, the size of the company, and so on. A maintainability engineer plays an important role in product design, with regard to the maintainability of the product. He serves as a specialist with respect to[12]

1. Maintainability program.
2. Maintainability analysis and prediction.
3. Maintainability trade-off studies.
4. Maintainability demonstration.
5. Maintainability data systems.

5.10. HUMAN FACTORS ENGINEER AND PRODUCT SAFETY DIRECTOR

This section briefly discusses the jobs of the human factors engineer and the product safety director. The human factor engineer plays a significant role during the system development. People who specialize in applied human engineering and human factors research have backgrounds in areas such as psychology, anthropometry, and physiology.[12] Ideally, the human factors engineer should be a "generalist" to achieve the best results. He must not be biased towards one particular discipline or one aspect of the system (e.g., controls and displays, physiology and psychology). Furthermore, in the system development, the human factor engineer must pay attention to the following:[13]

1. Arrangement of visual displays for optimum use.
2. Time sharing of two or more displays.
3. Functions to be accomplished to satisfy the system's purpose.
4. Proper utilization of decision-making and adaptive abilities of human beings.
5. Easy identification of each control device.
6. Environmental conditions with respect to human performance and physical well-being of individuals.
7. Overburdening the operators with information.
8. Adequate redundancy in the system for critical functions.
9. Arrangement of control devices for optimum use.
10. Appropriate grouping of various tasks into jobs.
11. Operational requirements of any given control within reasonable limits.

12. Compatibility of a control device operation with any corresponding display, as well as with expected human response.
13. Capability limits of human beings with respect to decisions.
14. Correct use of sensory modality for data to be presented by displays.
15. Use of correct display to present specified information.

A good human factors engineer understands and publicizes the following advantages of a human engineering program:[14]

1. It helps to reduce the training cost associated with a system.
2. It helps to increase the system reliability.
3. It helps to decrease losses from accidents and misuse.
4. It helps to improve user satisfaction with the equipment.
5. It helps to improve performance.
6. It helps to improve utilization of manpower.
7. It helps to have more economical production and maintenance.

The role of a product safety director is important with respect to product safety. That is, if the safety director fails to fulfil his responsibilities effectively, product safety may be sacrificed. Some of the responsibilities of a safety director are as follows:[15]

1. Analyze the historical data of product hazards, mishaps, and failures.
2. Keep the management informed of safety program performance on a continuous basis.
3. Make sure that the work is performed according to safety policies developed by the company.
4. Initiate proper measures so that mishaps do not reoccur.
5. Monitor the safety work being done by the subcontractors.
6. Provide safety-related information to those concerned.
7. Develop safety warning devices.
8. Establish guidance in executing a safety program from the design and development phase to the maintenance phase.
9. Establish criteria useful for analyzing any mishaps associated with a product produced by the company.
10. Perform analysis to make certain that product hazards are controlled or eliminated.
11. Audit all safety requirements that effect an item being manufactured, or to be manufactured by the firm.

5.11. PERSONNEL OFFICER

The position of the personnel officer is very important, whether it be in engineering or nonengineering organization, because directly or indirectly, day to day tasks are performed by people. People must be hired and proper care must be given to those already employed. (This is also true for reliability manpower.)

Some of the responsibilities of a personnel officer are as follows:[16]

1. Classify applications and interview applicants.
2. Evaluate qualifications.
3. Take part in policy making and overall planning to provide a uniform personnel services.
4. Look after administrative work involved with tests.
5. Confer with managers on personnel matters.
6. Coordinate personnel training activities.
7. Develop a service rating system.
8. Train supervisors to make evaluations of employees.

A personnel officer must have a number of years of experience in personnel administration. In addition, a degree in personnel administration is desirable.

5.12. SUMMARY

This chapter discusses the functions, qualifications, etc., of various engineering professionals, who are directly or indirectly responsible for product reliability and maintainability.

The chapter begins by discussing the characteristics and needs of an engineer, and is followed by a description of the activities and attributes of an engineering manager. In addition, the chapter outlines the basic qualifications of both the system engineer and the system project manager. Another topic discussed in the chapter is the objectives and authorities of a project engineer; namely, two basic objectives are listed, and eight areas of decision-making authority are presented.

Also included in this chapter are the characteristics of a design engineer, a number of guidelines for the design auditor, and the responsibilities and qualities of a configuration manager. In addition, the qualities of a contract negotiator are presented, as are a number of functions and attributes of a quality control manager. The responsibilities of quality control personnel are also outlined.

Reliability and maintainability engineers are discussed, and the functions and qualifications of human factor engineers are outlined. In addition, the responsibilities of the product safety director are presented. Finally, the chapter briefly discusses the responsibilities and qualifications of a personnel officer.

5.13. EXERCISES

1. What are the functions of a maintenance engineer?
2. What are the attributes of a good reliability and maintainability engineer?
3. Discuss the qualifications of a system safety engineer.
4. What are the responsibilities of a system safety engineer?
5. What are the qualities of a good contract engineer?
6. What are the similarities and differences between the functions of quality control and reliability engineers?
7. What are the essential qualifications of a design engineer?

5.14. REFERENCES

1. Randsepp, E. "The engineer—his intelligence and abilities." *Machine Design*, December 1959, pp. 29–31.
2. Couey, F. "What makes a good manager?" In *Management Guide for Engineers and Technical Administrators*, edited by N. P. Chironis, McGraw-Hill Book Company, New York, 1969, pp. 8.
3. Lessells, G. A. "Attributes a good manager should possess." In *Management Guide for Engineers and Technical Administrators*, edited by N. P. Chironis, McGraw-Hill Book Company, New York, 1969, pp. 9.
4. Chase, W. P. *Management of System Engineering*. John Wiley & Sons, New York, 1974, pp. 195–198.
5. Hajek, V. G. *Management of Engineering Projects*. McGraw-Hill Book Company, New York, pp. 1–5.
6. Harrisberger, L. *Engineersmanship: A Philosophy of Design*. Wadsworth Publishing Company, Belmont, California, 1966.
7. Landers, R. R. *Reliability and Product Assurance*. Prentice-Hall Inc., Englewood Cliffs, New Jersey, 1963, pp. 302–310.
8. Samaras, T. T., Czerwinski, F. L. *Fundamentals of Configurations Management*. John Wiley & Sons, Inc., 1971, pp. 34–35, 14–15.
9. Hicks, T. G. *Successful Engineering Management*. McGraw-Hill Book Company, New York, 1966.
10. Juran, J. M., Gryna, F. M., and Bingham, R. S. *Quality Control Handbook*. McGraw-Hill Book Company, 1974, pp. 7.1–7.2.
11. Hayes, G. E., Romig, H. G. *Modern Quality Control*. Collier Macmillan Publishers, London, 1977, pp. 447–448.
12. Woodson, W. E. *Human Factors Design Handbook*. McGraw-Hill Book Company, New York, pp. 16–17.
13. McCormick, E. J. *Human Factors Engineering*. McGraw-Hill Book Company, New York, pp. 608–610.
14. Chorafas, D. N. *An Introduction to Product Planning*. Cassell, London, 1967, pp. 192–193.
15. Kolb, J., Ross, S. S. *Product Safety and Liability*. McGraw-Hill Book Company, 1980, pp. 79–80.
16. Glueck, W. F. *Personnel: A Diagnostic Approach*. Business Publications, Inc., Plano, Texas, 75075, 1982, pp. 11.

CHAPTER 6

The Life Cycle Management of a System

6.1. INTRODUCTION

The life cycle of a system is the underlying basis for much of the discussion in this book. The life cycle of a system represents the various time connected phases, through which any system or equipment must pass. The management actions that must be taken during each phase are outlined.

6.2. THE LIFE CYCLE

All systems and equipment pass through four major phases between the time of their creation to the time of their disposal; reliability and maintainability management is involved in all phases. Prior to the life cycle phases, management is concerned with the initial policy planning and the initiation of follow-on planning, development, and definition. The four major phases of the life cycle are shown in Figure 6.1.

6.2.1. The Concept and Definition Phase

During the *concept* and *definition* phase a need for a product is established and the product's basic characteristics are defined. This activity is carried out through a process of documentation, which states the capability requirements and how they can be met.

6.2.2. The Acquisition Phase

The *acquisition* phase is the second phase of the life cycle. In this phase the management activities are directed toward product acquisition and installation, and toward planning for the eventual support of the system or the equipment chosen during the first phase.

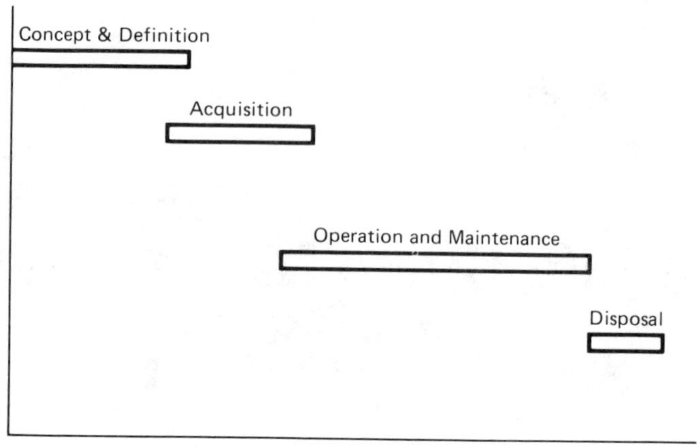

Figure 6.1. The life cycle.

6.2.3. The Operation and Maintenance Phase

The *Operation* and *Maintenance* Phase, sometimes called the *In-service* Phase, includes the management of the engineering, the maintenance, and the support of the system or the equipment during the entire operational life. This phase ensures that the capability requirements previously stated have been met, and will continue to be met, within the cost restraints.

6.2.4. The Disposal Phase

The *Disposal* Phase is the final phase and consolidates all tasks required to remove the system or the equipment, plus all its nonrequired supporting material.

6.3. MANAGEMENT TASKS

To achieve a desired level of reliability and maintainability during operation, a series of managerial tasks must be coordinated throughout all four phases. These management tasks, which are necessary for each phase, should constitute a management program. Choosing the appropriate tasks depends on the activity that the system or the equipment is engaged in. Some of the management tasks that must be performed during each phase of the life cycle are stated below.

6.3.1. Reliability and Maintainability Management Tasks in the Concept and Definition Phase

The reliability and maintainability tasks in the *Concept* and *Definition* Phase include:

1. Defining all terms. Agreed terms and definitions should be consistent throughout all life cycle phases. MIL-STD-721 and IEC-Publications 271 cover most terms needed.
2. Defining the system or equipment capability requirements. This definition should be in the form of a requirement specification which defines features and performance and the essential targets which must be met.
3. Defining the system or equipment concepts in terms of how these can be met.
4. Defining the system reliability and maintainability targets in quantitative terms, and defining a failure.
5. Defining constraints which management must deal with, such as time, cost, and manpower.
6. Defining the management controls which must be placed on both the designers and the manufacturers.
7. Defining the safety requirements of the system and the control of hazards.
8. Defining the parts control requirements and the methods to be followed in the design and the manufacturing phase.
9. Defining the basic maintenance approach and philosophy.
10. Defining the environmental factors under which the system or the equipment is manufactured, transported, installed, and operated.
11. Defining the hardware and the software standards to be followed, especially for the reliability and maintainability requirements.
12. Defining the constraints that have been found to be detrimental to reliability.
13. Defining the requirements for the collection and analysis of data during all phases.
14. Providing the appropriate documentation.
15. Providing the appropriate integrated logistic support system.
16. Defining the combination of machines, tools, facilities, and manpower required to produce the design and its assemblies to the stated specifications. The effect of these factors on reliability and maintainability should also be stated.
17. Defining the management controls required for documentation.

It is not necessary that all of the tasks listed be chosen in every case, because the choice will depend on the type of system or the type of equipment being used.

6.3.2. Reliability and Maintainability Management Tasks in the Acquisition Phase

The *Acquisition* phase deals with the actual acquisition of the system or the equipment; that is, any tasks that must be provided by management to obtain what is required. The system that has been obtained must be located or installed for operational use, and prepared for the next phase, namely, the *in-service* Phase. Also required is testing prior to getting the system or the equipment into service, and verification that the stated requirements have been met. The plan to provide for integrated logistic support, which has been defined in the concept and definition phase, must now be translated into the actual support needs. Similarly, many of the initial concepts must now be formulated to meet the acquisition requirements. More than in the first phase, both the customer and the contractor are involved, and both must be considered when defining the management tasks.

The key document, which management must prepare and which will lead to the final acquisition, is the *contract demand* or the *contract*. This document must spell out all the requirements that the customer wants from the contractor. It will include not only the technical requirements, the reliability and maintainability data, the support needs, the demonstrations, costs constraints, and warranties, but also the needed documentation and the controls. In addition, there are some legal, contractual clauses which must be considered, namely, how do these clauses affect the reliability and maintainability program. The acquisition phase must also plan how to train users and maintainers, and how to most economically support the system. Some of the tasks that lead to the acquisition, the installation, and the support planning are as follows:

1. Define the major design and development techniques to be used. For an off-the-shelf item, these two tasks are not required unless changes in the item are to be made.
2. Define all the technical requirements of the system.
3. Define all the reliability and maintainability requirements that must be met.
4. Define all the documents needed as part of the final system or the equipment.
5. Define what kind of prediction techniques should and will be used to assess the system.
6. Define the demonstration requirements, the tests, and the studies that

must be undertaken during design and development, and prior to delivery.

7. Define what is meant by a failure or a degradation and relate this information to the operational stated requirements.
8. Define what kind of future logistic support for a given period of initial acquisition and In-service time is required. This includes the facilities, the test equipment, and the spares.
9. Modification kits, if required.
10. Define the kind of reviews, such as design reviews, which will be held.
11. Define the type of installation, and define who is responsible for supplying the needed hardware.
12. Define what type of field studies, if any, should be conducted.
13. Define what kind of data the contractor must provide to the customer from the start to the end of the contract.
14. Define the controls that must be exercised by both the contractor and the customer during this phase. These will include configuration management controls, customer-contractor controls, audit controls, standard controls, etc.
15. Define the cost restraints and the life cycle cost information, which should be developed and considered for a given period of time.
16. Prepare the technical data package for procurement.

The impact these tasks have on the reliability and maintainability of a system or equipment is so great that management should evaluate each step carefully. Once the Acquisition Phase has passed and the In-Service Phase has started, much less can be done to further improve reliability or maintainability unless major design changes are made. This will be discussed later on. One area of concern to management should be the fast obsolescence of parts and material. Although standard parts are supposed to be available for a long time, continued changes in specifications, and low-quantity requirements force manufacturers to discontinue certain standard lines. This holds for both mechanical as well as electrical-electronic parts. The so-called theory of equivalency should be examined with great care. It is therefore important that the definition of spare parts requirements for initial or future needs be examined, keeping the above comments in mind. There is a direct cost relationship here that must be taken into account. Spending front-end money for all the spares needed for a life cycle of ten years or more is very costly and requires storing, accounting, and shelf life reviews. The question of shelf life is another aspect that must be considered, since possible deterioration could be a very costly part of the life cycle cost. Most of the military stores have systems in place whereby codes are assigned to shelf-prone items during the Acquisition Phase, which allows for controlled monitoring.

6.3.2.1. Acquisition Demonstration. One particular task is singled out not because others are less important, but because reliability and maintainability demonstrations leave much to be desired. An effective demonstration of these parameters is the responsibility that of management.

There are a number of constraints which are often used as excuses to cut demonstration requirements to a level such that any resulting information becomes meaningless for the purpose of decision-making. Some of the con straints that are put forward are (1) the schedule of fielding an item early; (2) the cost of any demonstration; (3) the unavailability of large quantities from production runs; (4) the required manpower to carry out the demonstration. The achievable level of confidence during the short demonstration cycle is usually not high, and managers sometimes feel that they do not gain much from such information. The first two constraints are usually the most significant from the manager's viewpoint. Systems with low failure rates, or very high MTBF (Mean Time Between Failure) values, cannot be fully demonstrated within the specified short demonstration period. The literature is full of various methods that can be used for reliability and maintainability demonstrations; some methods are reasonably simple and others more complex. But all attempt to demonstrate reliability and maintainability with maximum confidence and as economically as possible.

As suggested in Paragraph 6.3.2, managers must define (1) the prediction technique to be used; (2) the tests or analysis to be performed; (3) the kind of field environment the equipment should see; (4) the types of failures to be counted in relation to the operational requirements. Some useful methods are listed which are either normal demonstration tests or special tests, such as reliability growth. Chapter 11, Paragraph 11.4.4.4, deals with the data needs for reliability growth.

MIL-STD-757, Reliability Evaluation from Demonstration Data, shows how failure rate data can be used to arrive at a quantitative reliability assessment of an item.[12]

MIL-STD-781, Reliability Design Qualifications and Product Acceptance Tests: Exponential Distribution, is applicable to systems with Poisson failure distributions.[13] Sequential tests are easy to apply. Unfortunately, the tests have often been misapplied and unrealistically high estimates are advertised from such results.[4]

MIL-STD-471, Maintainability Verification, Demonstration Evaluation, includes steps which can be used to demonstrate the maintainability of an item.[14] It requires timing of induced faults through fault isolation, fix, and checkout. MTTR is measured against a minimum acceptable MTTR and a specified value.

IEC-56 (Secretary) 163, Draft Availability, Analysis and Demonstration, is a proposed guideline to carry out demonstrations of availability rather than reliability or maintainability alone. Many systems and equipments specify availability requirements.[5]

MIL-STD-1635, Reliability Growth Testing, describes the Duane model which is used to plot the reliability growth curve of the log MTBF versus the log operating time.[15] A variation of this model by Codier plots the data to aid in finding the number of burn-in cycles which are required to achieve a stated failure rate.[6] The Duane model takes into account total test time, the number of failures, the accumulated failure rate (a parameter dependent on design).

The Integrated Reliability Growth method uses fixing of reliability problems during the entire design/development cycle by collecting all failures and event data. This method allows collection of sufficient data to arrive at an early MTBF determination.[7]

The Large System Reliability Growth method models the following parameters: the initial reliability, the growth rate, the reliability to be achieved, the time it takes to isolate and verify failures and the resources required to continue plotting growth.[8]

The Reliability Growth Potential model expresses the expected reliability value from the experienced failure rates which have no corrective action or fixes and the percent decrease in failure rate after corrective action.[9]

The Reliability Growth Measure Demonstration model using AFM 66-1 Data makes use of long periods of data collection, sometimes over periods of up to ten years.[10]

The Bayesian Reliability Growth model assumes a prior density function to estimate MTBF values and their confidence limits periodically. A moving time window reduces the influence of early test data.[11]

The AMSAA (US Material Systems Analysis Activity) model, which is given in MIL-HDBK-189, plots the reliability improvement factor over a period of time in which changes in the product design or the manufacturing process occur. This model is more of a learning growth curve rather than a reliability growth demonstration method.[16]

Reliability demonstrations of software makes use of correcting faults in software by a continued reduction in the software failure rate, which leads to the specified design target value.

The preceding paragraphs indicated some of the methods that are available for demonstration purposes. The time that may be gained by eliminating demonstration will be felt further downstream in the life cycle and will aggravate the life-cycle cost, something no management likes.

6.3.2.2. Testability. As stated in Paragraph 6.3.2, the major design techniques and the technical requirements for system development must be defined by management during the Acquisition Phase. Especially critical is defining the testability requirements.

Due to increasing system complexity, system maintainers have been forced to demand from the manufacturer, a vastly improved method of identifying and isolating failures. The large number of no-fault-found maintenance activities, and the time-consuming failure isolation down to the module level, represents a prohibitive situation, which neither the maintainer nor management is willing to accept any longer. BIT and other test methodologies have helped to overcome many problems, but are only effective if testability requirements are designed into the system from the start. A method to assess the inherent testability of a proposed design early on, is important enough to be included in system design at a minimum redesign cost.

Operational reliability and mean time between maintenance (MTBM) are dependent on the effectiveness of the system testability. Therefore, the thoroughness, with which any testing methodology msut operate, must be addressed and defined.

It is impossible to fully automate a method that detects and isolates all failures. For example, a failure in front panel displays, by-pass capacitors, or pull-up resistors cannot be detected; therefore, values such as 98% or more for testability effectiveness must be questioned. Increase in BIT detectability requires additional circuits which, in turn, cause reduction in BIT reliability, even though the effect on the system reliability is negligible. Maintenance times can be reduced with increased testability effectiveness, but the number of maintenance actions will also increase. The increased number of maintenance actions is a penalty that must be taken into account when increasing testability thoroughness. In systems where security and safety are of prime importance, manufacturers have introduced BIT self-checking features; that is, BIT checking BIT.

Testability may be measured by relating the percent of failures detected or isolated, to the failure rates of the parts contributing to the failures (λ_c), divided by the total parts failure rate (λ_t), namely the failure rate ratio.

$$P = \left(1 - \frac{\lambda_c}{\lambda_t}\right) \times 100 \tag{6.1}$$

6.3.2.3. Sparing. Task #8 listed under Section 6.3.2, requires a definition of the spares. In Sections 7.3.9 and 7.4, the aspects of sparing and logistic support requirements are mentioned, and a few additional comments are made, which have a direct impact on the required management task during the Acquisition Phase.

The objective of sparing is to buy sufficient spare parts for entry into the inventory, which will be able to support the system during the life cycle. In most cases, the anticipated or expected life of the system is known. Life extensions may be envisaged, but usually this factor is not clear at the time of system introduction. For example, with semi-conductor devices often becoming unavailable after only short time periods, some consideration must be given to how many spares should be purchased initially. The cost of a life-time buy may not be feasible, however, due to limited available funds. In such cases, initial purchase of a limited number of spares is likely to encounter the problem of eventual parts replacement, and this situation will be aggravated when life extensions are being planned at later stages.

Where should the data for calculating how many spares are needed come from? Estimation of any required quantity for an initial buy, a life-time buy, or a life-extension buy should be based on the parts failure rate and not on the experienced inventory demands of the same or a similar item. Inventory demands do not reflect the failure rate of parts for many reasons. If the failure rate of the required part is available, this data should be used. New parts without an established failure rate should make use of MIL-HDBK-217 models, or other failure rate models. A problem may arise, however, when a part represents a common user item; that is, it is used in a number of systems. Since each system may have a different application, the part may experience different failure rates in each. Usually, such parts are binned together as a common user item, and care must be taken that the requirements of all systems are met when deciding on the quantity to buy for a particular system.

6.3.3. Reliability and Maintainability Tasks in the In-Service Phase

This phase, also known as the Operation and Maintenance Phase, ensures that the appropriate steps are taken to keep the system or the equipment maintained and supported during the entire operational life period, and that the stated requirements have been met and continue to be met, within the cost limitations imposed. The management responsibilities can be divided among three distinct engineering groups: (1) the actual user, the actual operator, and the maintainer; with (2) the material manager during the life cycle of the system; (3) the supply manager. The last two groups, in many cases, integrate their functions. The reliability and maintainability management is the responsibility of the second group. The tasks to be considered may be described as follows:

1. Developing engineering and maintenance documents. These may be policy, procedure, instruction, guidance, or engineering documents. The preparation of such a large number of documents requires a substantial management effort, especially if these are to be effective within the time frame to which they apply. Too often, development

takes place after the fact, thus making the documents ineffective because they lag far behind the actual conditions to which they apply. These documents become the corner stones for the users, the maintainers, the suppliers, the trainers, the life cycle material managers, and any other decision-making body.

2. Management control of all the documents. Configuration management in the In-Service Phase is very critical. Without this control over the hardware/software, changes in reliability and maintainability, analysis, audits, or reviews are almost impossible. Configuration management takes into account the total system rather than just a single item.
3. Collecting, monitoring, and analyzing reliability and maintainability data. The purpose is to (1) identify changes from the predicted values; (2) alter, if required, the support needs; (3) carry out trend analysis; (4) determine causes of problems; (5) institute corrective actions; (6) prepare changes to documentation; (7) assist in the design and decision-making processes.
4. Establishing data banks. The availability of data and the proper monitoring of data can produce information at all levels of the system or the equipment, and can assist in evaluating new systems, off-the-shelf systems, and in-service systems. The collection and analysis of data from a data bank is essential to any in-service reliability and main tainability program.
5. Managing and predicting spare parts. Managing the purchasing, the storing, the distributing, the accounting, and the inventory controlling of spares is the responsibility of the supply-type organization. Prediction and rationalization is a function of the life cycle manager. Logistic planning ensures the availability of spares to replace failed items, to minimize the inventory, and to centralize activities, if possible. To prediction the number of spares required involves the use of analytical tools, which are described in another chapter.
6. Providing maintenance tools and test equipment. This involves the definition of tools and test equipment for maintenance, and the development of plans for their appropriate location and availability when required. Test equipment can be a very costly part of the life cycle cost. System availability and turn-around factors play a major part when assessing these support resources.
7. Providing adequate manpower. Large systems may require trained manpower which is not readily available. A plan must be devised to train the needed manpower for operating, controlling, diagnosing, maintaining, and supporting the system. Both hardware and software training may be needed because many modern systems are software-controlled.

8. Developing maintenance support for the various maintenance levels. Third-line maintenance support requires management action because it is an outside activity, not under direct control of the user or the maintainer. Actions arising from this support line have a direct influence on reliability and maintainability, and system availability to the user.
9. Preparing Engineering Change Proposals. Systems that have been introduced into an inventory, may, after a certain time, no longer meet all the requirements for operation, because of operational changes or decreasing reliability and increasing maintainability. In such cases, engineering changes may be required. A management decision is needed to weigh the cost of replacement against the possible life extension, or the modification of the system.
10. Reviewing the documentation, procedures, instructions, and support requirements in light of any engineering change. (Changes in reliability and maintainability must be taken into account.)

The above tasks are some of the major tasks that must be performed during the In-service phase. Depending on the type of system or equipment, other tasks may have to be added.

6.3.3.1. Parts Replacement Task. A fairly recently required task is parts replacement. Solid state devices such as ICs, have proliferated the market to such an extent that volumes of IC listings have been created. Not only is there an explosion of specially designed ICs, but each manufacturer incorporates certain features in his IC design to make them unique in character and packaging. Standardization has been somewhat limited to a few devices, but a standard approach of testing ICs has at least been adopted by the majority of manufacturers and customers (MIL-STD-883, Test Methods and Procedures for Microelectronics).[17]

Initially, most devices were manufactured in sufficient quantity to satisfy the demand. But unless large enough reorder quantities were placed, many ICs were no longer produced, and were taken off the market. Large expenditures must be justified to remanufacture certain parts, in smaller quantities, and for specific requirements, otherwise managers are faced with the problem of parts replacement. On the average, the market life of ICs is now in the order of a five-year life span, but many systems and equipment have life expectancies of ten or more years. An initial purchase of sufficient ICs for the total life cycle of a system is one possible solution to the parts replacement problem, although such a purchase involves a high initial cost factor, and the risk of buying either too few or too many of a particular part. (This is due to the lack of available information on actual failure rates on which spare quantities purchases can be based.) Extended shelf life for ICs is another factor that influences reliability, as has been clearly demonstrated on some long dormant missile systems.

It is anticipated that for the time being there will be a continuous increase in IC parts replacement requirements, and it is advisable for management to give thought to developing a task plan or a guide which can be used for parts replacement. (Such a plan should be flexible enough to allow any type of IC part replacement.) A number of steps should be taken when considering such a task plan. The first step must be the identification of the problem. Here, the question is, Is the required device available from any stock? This question requires that a search task be performed. If the answer to the question is no, then the original part cannot be replaced by purchase from stocks. The next steps is the characterization of the device, the chip, and its package, as well as its application. This detailed documentation is essential for any further action. The availability of both a suitable chip and the package must then be determined. (Unless all this information is readily available, this task involves expert knowledge of IC designs.)

If a chip is available, it may have to be repackaged if the original package is no longer manufactured. If the chip is not available from any source, then a new chip must be designed, or an existing chip must be programmed. In any case, there are a number of different approaches to be used, and management must be advised by an expert as to which of all possible solutions may be best. For example the solution might be to design a linear, a gate array, or a programmable device configuration. The best choice will depend not only on technical considerations, but on the required quantity and the cost factor. Once a decision has been made, a manufacturing, design, development, or packaging source must be selected. Next, a contract data package must be prepared, which includes all the required details, such as testing, burn-in, and accept/reject criteria.

6.3.4. Reliability and Maintainability Management Tasks in the Disposal Phase

The disposal of a system may result from a number of situations: (1) the system may be obsolete and unable to fill the role that the user wants; (2) the system may be unreliable, and thus, rising support costs may make the system uneconomical; (3) the system may not be compatible with other systems with which it has to operate; (4) the sale of the system to another user may have a cost advantage which cannot be obtained at a later stage; (5) functional replacement exists in the inventory, and thus, the system becomes a burden to inventory. With few exceptions, the question of reliability or maintainability becomes an issue in the Disposal Phase. (One exception might be if the system is sold to another user.)

Many tasks may have to be performed during the Disposal Phase, which have no impact on reliability or maintainability. There is one area that may have to be considered. One such task might be the identification of the number of common user items, which are part of the system to be disposed of. Therefore, some of the management tasks that might apply are:

1. The identification of any common user item that the system has been using.
2. The calculation of the usage rates of these common user items, to determine the number of items that can be disposed of from the total common user stock, and which cover only the requirements for the system to be disposed of. Failure rates encountered during the In-Service Phase may provide some quantitative values.
3. The calculation of the final reliability and maintainability values of the system. The buyer of the system requires this for his operation. This information must be useful for future actions on similar systems and should be stored in a data bank.
4. The final life cycle cost should be calculated, which includes the disposal action and cost recovery, if applicable.

6.4. LIFE CYCLE MANAGEMENT CONSTRAINTS

Like any management system, life cycle management has built in certain constraints and limits. If an organization wants to adopt the life cycle management system, and specifically deal with reliability and maintainability management, certain facts should be considered. For example, a system that requires a large number of tasks over the life cycle may appear simple on the surface, but its degree of complexity cannot be accurately ascertained by merely adding up the various tasks. The tasks are often overlapping; that is, as one task starts, another must, at the same time, be considered. For example, in the Concept and Definition Phase, the task of defining the requirements for the collection and analysis of reliability and maintainability data must be closely coupled to the task of defining the necessary documentation requirements, as well as the kind of management control for these documents. Many more examples could be shown, although it should be obvious that the interrelation between various tasks makes it necessary to consider all tasks collectively, before deciding how to make the management system work effectively for the organization.

The entry of the system is another aspect in the timing period of the total program. At the start, when only general concepts of a program are being discussed, the system may not be of major concern until more detailed questions are answered. The overall operational or user requirements, which are tabled before any technical details are discussed, do not, in most cases, require the presence of a life cycle material manager. There will be some overlapping, when the various parties meet, and it is at that stage that the topic of the system enters the discussion.

6.4.1. Delegation of Tasks

Since the total life cycle of a system or equipment may extend over a very long period of time, it is almost impossible for one individual to remain in a life cycle management position for the entire life cycle. In addition, it may be impossible for one individual to be familiar with all technical aspects of a project. Reliability and maintainability is a subject that requires specialized training and the manager may not be an expert in this field. He will have to delegate this function, and possibly some others, to more experienced professionals, who in turn report their findings and recommendations to him (as the senior manager of a project). Engineering decisions that influence reliability or maintainability may have to be guided by delegated experts. Any maintenance-related decisions may be made directly in the field, the repair shop, or the repair depot, and the manager must be made aware of all of them. If the manager is not aware of all maintenance-related decisions, he loses the needed control, and configuration management becomes a slogan rather than a management tool. The responsibility for configuration management may lie with a contractor, especially during the overhaul and repair cycle, as he is more familiar with the details of any modifications, changes, or configurations than anyone else.

6.4.2. Funding

Over the life cycle, the manager will be faced with a number of major and minor expenditures. Reliability and maintainability will affect the cost not only during the acquisition phase, but all the way downstream to the final disposal action, with the major funding during the In-service Phase. Does the life cycle material manager have control of the funds required for this period? In the majority of cases, no. Although this situation may appear strange, the material manager may request funds, assign funds for certain needs, forecast his funding requirements by estimates, and prepare documents that require expenditures (e.g. contracts). Who decides what funds will be provided for a project? Who signs all funding expenditures? Who controls all funds? Who can increase or decrease funds? In practically all cases, the answer is not the life cycle material manager, but rather someone else in the finance office—the financial manager. (This is especially true for larger organizations and the government.)

A great deal of interfacing is required which may be viewed as a constraint factor. The objectives are to acquire, operate, maintain, and dispose of material as economically and effectively as possible. These activities demand interfacing with operators or users, financial organizations, supply people, and specialists in fields such as reliability, maintainability, life cycle costing, configuration management, standardization, documentation, and software engineering. The art of delegation of subtasks is an important factor for a successful life cycle material management system.

6.5. SUMMARY

The various life cycle phases have been described together with the actions that management must take and is responsible for. Such actions ensure that the final product meets the intended reliability and maintainability within the cost constraint.

6.6. EXERCISES

1. Briefly explain the four life cycle phases, and list each of the specific activities that have a direct influence on reliability and maintainability.
2. The life cycle material manager has many tasks to perform. What kind of organization should he have at his disposal? Prepare a simple organization chart showing the various functions that must be performed by various personnel.
3. What kind of actions are initially required to spare for a specified reliability of a piece of equipment? List some of the input factors which must be available for this purpose.
4. Corrective and preventive maintenance actions are the responsibility of the life cycle Material Manager. Express in words, what both terms mean, and how these can be measured quantitatively.
5. A maintenance management information system is required. What are the data that must be collected for reliability and maintainability?
6. Describe some of the constraints that the life cycle management system must deal with.

6.7. REFERENCES

1. Arsenault, J. E., Roberts, J. A. "Reliability Management," by A. P. Harris In *Reliability and Maintainability of Electronic Systems,* Computer Science Press, 1980, pp. 24–37.
2. Reiche, H. "Reliability and Maintainability as Part of Life Cycle Management," In *Proceedings National Defense Engineering Symposium,* Canadian Forces Base Trenton, October 1974.
3. Blanchard, B. S. *Logistics Engineering and Management.* Prentice Hall Inc., 1981, pp. 312–348.

4. Elliot, T. W. "Common Errors in Application of MIL-STD-781." In *Proceedings Annual RAMS Symposium,* 1984, pp. 139–142.
5. American National Standards Institute. International Electrotechnical Commission, Guidelines for Reliability and Maintainability Management, 56 (Central Office) 98, 1984. New York, N.Y.
6. Codier, E. O. "Reliability Growth in Real Life," In *Proceedings Annual RAMS Symposium,* 1968, pp. 41–53.
7. Bentz, R. W., Hutchinson, Maj. L. T. "An Approach to Reliability Growth Testing Without Dedicated Testing," In *Proceedings Annual RAMS Symposium,* 1984, pp. 458–464.
8. Smith, D. K. "Planning Large System Reliability Growth Tests," In *Proceedings Annual RAMS Symposium,* 1984, pp. 471–475.
9. Crow, L. K. "Methods for Assessing Reliability Growth Potential," In Proceedings Annual RAMS Symposium, 1984, pp. 484–489.
10. Boiles, J. H., Hadel, J. J. "Reliability Growth Measure by AFM 66-1 Data," In *Proceedings Annual RAMS Symposium,* 1981.
11. Tung, S. S. "A Reliability Growth Model," In *Proceedings Annual RAMS Symposium,* 1984, pp. 490–492.
12. Naval Publications and Forms Centre. MIL-STD-757, *Reliability Evaluation from Demonstration Data.* Philadelphia, PA.
13. The Naval Publications and Forms Centre. MIL-STD-781, *Reliability Design Qualification and Product Accept Tests.* Philadelphia, PA.
14. Naval Publications and Forms Centre. MIL-STD-471, *Maintainability Verification, Demonstration Evaluation.* Philadelphia, PA.
15. Naval Publications and Forms Centre. MIL-STD-1635, *Reliability Growth Testing.* Philadelphia, PA.
16. Naval Publications and Forms Centre. US MIL-HDBK-189, *Reliability Growth Management.* Philadelphia, PA.
17. Naval Publication and Forms Centre. MIL-STD-883, *Test Methods and Procedures for Microelectronics.* Philadelphia, PA.

CHAPTER 7

Activities of Reliability and Maintainability Management

7.1. INTRODUCTION

A number of different managers must be responsible for all reliability and maintainability activities, which must be carried out during the life cycle phases. As systems or equipment go through the life cycle phases, different managers will be involved in decision-making and management.

Each life cycle phase may have different management activities at various levels. The titles chosen for the managers involved with these required activities have been taken from practical experiences in government and industrial concerns, and reflect the functional activities of each manager. Three types of managers have been identified: the project manager, the life cycle material manager, and the supply manager.

7.2. PROJECT MANAGER

During the Concept and Definition Phase the Project Manager is responsible for a number of tasks directly related to reliability and maintainability. The performance achieved during the In-Service Phase can be no better than that specified during the Concept and Definition Phase. The specification created during this phase is the keystone upon which all subsequent activities are based. The specification must not only contain the intended performance requirement, but must define the reliability and maintainability targets for the benefits of design and manufacturing. It must also include the quantitative requirements. Because the specification is a basis for contractual action, its intent must be clear if a suitable product is to be obtained.

7.2.1. Terms and Definitions

One of the first responsibilities of the project manager is to define the terms and the associated definitions which will be used during this project. A number of documents exist which define reliability and maintainability terms. All differ slightly from each other, and a choice must be made as to which terms apply to the project. Otherwise, each person dealing with the project may have different views and interpret terms so that they can be used to their advantage. Such differing interpretations often lead to contractual problems later on, which can be rather costly.

One area that requires a clear definition is What constitutes corrective and preventive maintenance? Project management must state what aspects of maintenance actions are attributable to one or the other. The inclusion of related maintenance workload into preventive or corrective maintenance timing factors can alter quantitative requirements considerably.

7.2.2. Documentation for Project Management

The preparation of the requirement specification is one of the tasks for the project manager. This document identifies the requirements during the Concept and Definition Phase, and should list tasks appropriate to the type system or equipment to be managed. The priorities of these tasks must be identified to allow the project manager to better assess the possible trade-offs, and the level to which each task is addressed. The time frame for the tasks may be required. A few paragraphs are required listing the various information documents which must be raised for a particular project. (This information may serve a future role in major modifications.) A brief description of the envisaged role and function of the equipment is needed, and identification of major technical characteristics may be listed; this information will be used by the managers.

7.2.3. Performance Requirements

Since performance is closely related to reliability, the capability characteristic must be stated in some form, even though this information may only be available when a certain level of reliability has been reached. Consideration must be given to operating the equipment continuously, periodically, for a short time, or only once. Capability descriptions may vary from a simple statement to a lengthy statement. Multi-mission capability will require more extensive description; that is for multi-mission or multi-function equipment, a matrix of capabilities versus priorities may be useful especially when certain capabilities are more important than others.

Performance capabilities for equipment or systems are as follows:

- Receive.
- Transmit.
- Store.
- Detect.
- Search.
- Protect.
- Generate.
- Switch.
- Maintain.
- Test.
- Calibrate.
- Display.

These capabilities affect not only the performance of a system or equipment, but also their reliability and maintainability. The project manager must therefore define the reliability and maintainability targets.

The final product depends on the accuracy of statements which are made, and these statements will find repetition later on in the testing, the training, the analysis, all of the documents, and any other studies. The requirements can only be prepared at this stage by the designers, the developers, the maintainers, the logistic supporters, as well as the final user. These various managers can state the limits which they believe are necessary to a project or to resolve problems. The definition of the effectiveness of the system or equipment is a management function, and optimizing their design makes reliability and maintainability prime factors which must be integrated with other design criteria. The degree to which management emphasizes reliability and maintainability during the initial part of the life cycle will become evident in the achieved performance.

7.2.4. Specification Targets

The requirements or targets specified by the project manager must include:

- Quantitative terms, such as MTBF (Mean Time Between Failure), MTTR (Mean Time To Repair), or other factors.
- A time frame, because such quantities do not hold for the entire life cycle.
- A definition of a failure.
- An apportionment of values down to the lowest self-contained level.
- An essential and a desirable quantitative value.
- A minimum acceptable level for an initial equipment or system under certain conditions.

- An indication of how these stated values are to be determined by testing, analysis, or experiences.
- A type of support system which meets the maintainability values quantitatively.
- An indication of interchangeability, modularization, and fail-safe features.
- The requirements for BITE and testability.
- The software requirements.
- Needs for special training, tools, and facilities.
- Safety features.
- Configuration controls.

7.2.5. Maintenance Philosophy

At this early stage of the project, project managers must make certain decisions regarding the maintenance philosophy. The extent and timing of such decisions has a major bearing on system or equipment readiness, as well as cost. Lack of quantitative maintenance values may require assumptions regarding tasks required, and analyzing basic features of the planned design. A fundamental philosophy is that the magnitude of maintenance times for a discrete maintenance action, is the sum of the individual maintenance tasks required for completion. The major tasks can be assumed to be

- Fault localization.
- Fault isolation.
- Disassembly.
- Interchange.
- Reassembly.
- Alignment.
- Check-out.

Maintenance philosophy assumes that each of the maintenance tasks times is a function of the method of maintenance and the level in which the maintenance is performed. Functional levels may be described as

- System level.
- Subsystem level.
- Equipment level.
- Group level.
- Unit level.
- Assembly level.
- Subassembly level.
- Part level.

Project management must consider the different maintenance organizations and their concepts. Three major levels of maintenance can be identified. The first is the field maintenance or the maintenance at the site of installation. This type of maintenance can be carried out at the first line maintenance and is limited. The second line of maintenance normally consists of a small workshop, a repair shop, and a mobile test facility. More sophisticated test equipment and tooling is available, which allows more in depth maintenance, although more training is required for this line. Third line maintenance is normally the repair depot, the contractor, or the manufacturer, through which all maintainable items can be repaired and maintained. These three maintenance levels influence the entire system or equipment, and thus, compromises of maintainability must be evaluated, documented, and reflected in the total program (especially when dealing with large complex requirements.).

7.2.6. Procedural Requirements

The cornerstone of any reliability and maintainability program is a thorough preparation of the specifications and the procedural requirements from the design to the in-use stages. Contracts are often handled by persons who overlook the importance of specifications, procedures, and regulations. They do not realize that the contents of such documents are to be considered as part of the contract. One specification or standard contract clause may invoke other specifications, which in turn invoke more, some contradicting each other. If nothing is mentioned, however, it must be assumed that all referred documents, directly and indirectly, in full, are applicable. Unless a specific issue of a document is to be applicable, the issue in effect on the date of the offer for tender is assumed. Conflicting requirements may be costly and seriously affect reliability.

Consideration must be given if the procedural requirements include demonstrated, analyzed, or projected quantitative values. Demonstrated reliability and maintainability requirements may command premium prices, or may not be applicable to the small, anticipated production quantities, at the time of delivery. As part of any reliability management control a number of equipment subsets can be brought to bear on trying to achieve the stated requirements as early as possible. Such subsets include burn-in screens and growth management programs, and should not be regarded as separate management functions but as part of the total management program.

7.2.7. Specification Constraints

No system or equipment operates without some constraints, which may effect operational performance, engineering, and logistic support factors, as defined by management. Operational constraints directly influence the operational characteristic of the system, and the logistic support constraints have a major impact on maintainability and system readiness. Types of logistic constraints may be

- Limited availability of spares.
- Distribution of spares.
- Availability of support personnel.
- Skill of support personnel.
- Availability of support facilities.
- Availability of software.
- Availability of transport.
- Availability of test equipment.
- Funds.

One major constraint in electronic equipment is the limited manufacturing life of solid state devices, which in most cases, is less than the expected equipment life. Available funds and deterioration due to long shelf life may prohibit full lifetime spares. Thus, this constraint must be addressed by management.

7.2.8. Field Studies

Field study may be required for various reasons and have one common aim—the collection of information which can be analyzed for decision-making purposes. Depending on the intent of the study, tasks must be identified for the study, and the central responsibilities of the study must be shown. The project manager should address the following:

- Aim of the study.
- Intent of the study.
- Description of the item to be studied.
- A time table.
- Organization responsible for the study.
- Type of personnel involved.
- Study environment.
- Facilities required.
- Funding.
- Priority.
- Type of reporting system.

The reliability and maintainability parameters for the field study should be presented in quantitative statements and graphs to allow easy interpretation by management. A statement of confidence should be included in the information being presented.

7.2.9. Initial Documentation

The required operating and maintenance documents must be defined. During the design stage some of these documents may be very general until such a design has been formalized. Detailed documentation of the system by the life cycle manager will follow later once firm hardware is manufactured and eventually turned over to the user.

7.2.10. Logistic Support

Another task for the project manager is to define logistic support. A number of factors contribute to logistic support, and all influence system effectiveness. The integration of these factors at the start of the project, into a single management program is essential. Logistic support is initiated early in the conceptional stage, and is used by the contractor to develop the proposed logistics support requirements. More information on logistics support management is covered in a special chapter on this subject.

7.2.11 Other Management Controls

A number of other management controls must be established between the customer and the supplier. A review of these controls by both parties is required to ensure that they are agreed upon. It is important that only those requirements considered essential to the stated objective, and which can be enforced are specified.

Request for plans and programs from the supplier, prior to the design or the manufacturing, should be tabled. The degree of disclosure and the appropriate safeguards for the commercial sensitive information must be reviewed. The participation by the customer in design reviews has been mentioned in the special chapter on Design Reviews. The request for reliability demonstration by testing during the design and manufacturing stages has been mentioned. The requirement for audit by the customer should be included in the relevant task descriptions. The use of warranties will be discussed in a separate chapter. The document and software control in the form of configuration management will be discussed in another chapter.

7.2.12. Summary

The functions, actions, and responsibilities of the project manager, whose actions set the stage for the life cycle manager, have been identified. The next section deals with the life cycle manager.

7.3. THE LIFE CYCLE MATERIAL MANAGER

The life cycle material manager (sometimes called the material manager) is responsible for guiding the activities necessary to fulfill system or equipment management objectives. The management objectives are to obtain, operate, maintain, and dispose of material as economically and effectively as possible. The activities necessary to fulfill the management objectives are engineering and maintenance management, with a great deal of interfacing with the project and supply managers. Moreover, some interfacing must take place with the operations, the finance, and the personnel departments. The transfer of responsibilities from the project manager to the life cycle material manager occurs normally near the end of the Acquisition Phase.

7.3.1. Activities

Although the number of activities involved in system or equipment development is great, all have a major affect on reliability and maintainability. The various activities will be discussed in some detail. The following activities are tabled at random, and in no way reflect the order in which they are dealt with by the manager.

- Engineering services.
- Scientific and engineering information.
- Corrective and preventive maintenance engineering.
- Integrated logistic support engineering and management.
- Configuration management.
- Life cycle costing.
- Provisioning.
- Modification.
- Repair and overhaul.
- Quality assurance.
- Material storage conditions.
- Test equipment.
- Contract warranties.
- Specifications and standards.
- Technical data packages.
- Material Authorization.
- Technical orders.
- Contracts.
- Engineering Drawings.
- Specialty training.
- Budget and funding requirements.

7.3.2. Engineering Services

Since the manager may not always be familiar with reliability and maintainability engineering and management techniques, it may be necessary to obtain assistance from other experts. Engineering services may be used to provide specific services, such as engineering investigations and studies, and technical support for projects that require reliability and maintainability work. Normally, such services involve a single project with starting and completion dates specified. The selection of the best available engineering service is critical, and although a large number of consultants advertise reliability and maintainability engineering and management services, often the most valuable service uses an original, rather than a simple textbook approach. (Engineering service contracts should make this clear, when soliciting for services.)

7.3.3. Scientific and Engineering Information

The manager must have available to him, the scientific and the engineering information related to reliability and maintainability. He requires a wide range of information for the various projects—from simple maintenance data, to research, development, design, production, and operation data, to life cycle material management data. The number of information sources that can apply to reliability and maintainability are endless. For each particular job which the manager has to deal with, the same or different information is often required. It is therefore essential that he has available to him, the kind of specific information he requires. The sources of this information may be either external or internal; both will be discussed.

7.3.3.1. Internal Information. Information concerning the system or equipment operation and failure enables the manager to identify problem areas, and implement corrective action. An information system that provides reliability, maintainability, and logistics parameters is a necessary support feature for life cycle material managers. The system must be designed to provide information that highlights these parameters and problems which require management investigation and corrective action. Such information is essential to support the life cycle material managers' recommendations for the initiation of investigations, modifications, or system or equipment replacement.

The manager must either develop and implement an internal information system, or, if such a system is in existence, he must make himself familiar with the system. He must understand the capabilities and limitations of such systems and make use of the system data banks. Unfortunately, there is no single system that can give him all the information. Moreover, the data collected is not always correct, especially if it comes directly from a field operation where operators and maintainers are busy trying to get failed items back to the field as fast as possible; that is, the life cycle managers like to have as much information available to them as possible. On the other hand, the operators and maintainers like to reduce the input to data forms. The life cycle manager must decide what is the absolute minimum information with which he can effectively operate. Even with the best information system design, it is essential that he interface with other systems in order to complement his data. Chapter 11 discusses in more detail some of the data needs.

7.3.3.2. Existing Internal Information Systems. Many large internal information systems exist, which deal with reliability and maintainability. Most of them were developed for military users. A few examples of such systems as follows:

Canada—National Defense:

- Ship Maintenance Management Information System.
- Land Ordinance Maintenance Management Information System.
- Aircraft Maintenance Management Information System.
- Electronic Maintenance Management Information System
- Test Equipment Maintenance Management Information System.

US—Department of Defense:

- USAF — 66-1.
- USN-Aviation — 3-M (Air).
- USN-Ships — 3-M (Sea).
- US Army — TAMMS-SDC Army Maintenance Management System.
- US Army — System 2000.
- US Army — MANAGE System.

UK

- RAF/RN — Defect Reporting Systems 760 and 761.

7.3.3.3 External Information Systems. One of the best known external information systems in the United States Government funded Government-Industry Data Exchange Program (GIDEP), designed to assist in the dissemination of reliability and maintainability engineering and management information. Data interchanges cover engineering, metrology, reliability, maintainability, and failure experience from government and industry. This extensive automated data bank is a major source for all managers. The selection of information from this data bank can considerably enhance the effectiveness of system and equipment management.

(MIL-HDBK-217)—a military handbook which is updated continuously. Contains reliability information on practically all electronic parts. Automated data banks of this handbook make it possible to evaluate failure rates of parts within a short time. This kind of information is essential for parts provisioning using mathematical models, for reliability evaluations of systems and equipment, for maintenance planning, and for many other management tasks. A similar handbook is now being prepared for mechanical parts. Rome Air Development Center N.Y. issues a large number of documents dealing specifically with the reliability of semiconductors. This is a vital source for such data which is otherwise hard to come by. Dissemination of all information required for a particular project from internal or external sources is a task which is the responsibility of the life cycle material manager.

7.3.4. Corrective and Preventive Maintenance Engineering

Corrective and preventive maintenance is sometimes called unscheduled and scheduled maintenance. When, however, the maintenance is for aircraft and related equipment, the terms *primary* and *periodic inspection* have been used instead of preventive maintenance. The function of the life cycle material manager is to develop maintenance procedures and schedules that assure high availability, which is the main objective of all maintenance activities.

Corrective and preventive maintenance information must be published for systems and equipment. For complex, new equipment preventive maintenance schedules and procedures can be established only from information generated by the manufacturer or prime contractor, or after extensive field experience. (The maintenance management information systems provides data for this purpose.) Failure trends can be used to modify and improve the maintenance information. The extension of the intervals between preventive maintenance is a task which must be studied carefully. The increase in number of parts tends to require an increase in maintenance activities, but conversely, parts have become much more reliable with reduced failure rates. Changes to the interval should be planned through studies and field trials.

7.3.5. Integrated Logistic Support Engineering and Management

The application of an *integrated logistic support plan* should be considered by the manager for any major system project. The range of options is large, and decisions by management may substantially improve reliability and maintainability and realize life cycle cost savings. A plan may be prepared by a contractor in response to a statement of work, or may be produced by the project manager. The life cycle material manager will be responsible for implementing the various engineering and management functions during the In-Service Phase of the life cycle. The monitoring of the effectiveness of the plan during this phase is a major requirement. The lack of any problems during the In-Service Phase is an indication of a good integrated logistic support plan.

7.3.6. Applicable Documents

Two documents deal specifically with engineering and management support and relate directly to reliability and maintainability. MIL-STD-1388[1] deals with Logistic Support Analysis and MIL-STD-1369[2] with Integrated Logistic Support Program Requirements. These documents contain some of the following support requirements:

- Maintenance planning.
- Support and test equipment.
- Supply support.
- Transportation.
- Technical data.
- Facilities.
- Personnel and training.
- Broad maintenance concept.
- Maintenance planning factors.
- Initial reliability and maintainability parameters.
- Initial repair levels.

Logistic analysis, vital to the development of any maintenance plan, identifies the specific logistic support elements required. This analysis begins during the Concept Phase. The analysis also provides data for repair and for optimum repair levels, and examines the available repair options for each configuration. Of interest are repair frequency, average repair times, quantities and cost, publications, tools, test equipment, transport, and personnel. All of these factors have a direct affect on system availability management or system effectiveness. MIL-STD-1369 should be used to develop the procedures and format for an integrated logistic system program. Both documents are designed for input to an automated data program, and this should be kept in mind by the manager who uses these documents.

7.3.7. Configuration Management

It is the responsibility of the life cycle material manager to ensure that all specified system or equipment characteristics are accurately described in the engineering documents. Configuration management begins with the specification of a base line configuration. All elements that make up a system must be specified at the start. Configuration identification applies to the content of approved technical documentation, and alternations of the configuration are normally called *engineering changes*. The manager is responsible for approving, evaluating, and coordinating such engineering change proposals. Changes may affect performance, reliability, maintainability, cost, or a number of other factors. The military divides these changes into two classes which are outlined in detail in DOD-STD-480.[3] When a system or equipment is modified, a decision to reidentify the item may be the result of a recommendation by the life cycle material manager or the supply manager, as these activities are part of configuration control and management. Another aspect of configuration management deals with any design change deviations, such as waivers. Configuration management, which constitutes a vital management task, has been defined as "a discipline applying technical and administrative direction and surveillance to identify and document the functional and physical characteristics of a configuration item, control changes to those characteristics, and record and report change processing and implementation status."

7.3.8. Life Cycle Costing

True life cycle cost can be determined only after a system or equipment has been withdrawn from service, at which time the total expenditures for hardware, software, personnel, operations, and maintenance are added up. To be an effective management tool, true life cycle cost analysis must be based on realistic estimates of the cost of ownership. Life cycle cost estimtes are used to select the most economical system available for a required task or role. In addition this estimate is used to evaluate the cost of retaining an item for some additional time period, as opposed to replacing an item. Life cycle cost implications should be studied carefully by the life cycle material managers whenever a decision is made that will affect operation, maintenance, and support of a system or equipment. When making such a decision, trade-offs must be considered and alternatives must be evaluated. The mathematical calculations required for such studies are described in another chapter.

7.3.9. Provisioning

The life cycle managers are responsible for technically assisting a provisioning team which selects the items that must be procured to support a system or equipment. To assist such a team, the manager must be familiar with the item prior to the selection task, and must participate in the examination of the documentation which identifies the reliability requirements for an item. The provisioning of items with reliability requirements requires knowledge of certain reliability information, and methods to arrive at decisions regarding the quantity of spares to be purchased within a given budget. The quantity of consumable items should be sufficient to last for a period of three to five years, depending on lead time, unit cost, economic order quantity, and shelf life. The type of documentation required for the purchase of reliable spare parts should consist of long lead lists, recommended spare parts lists, a provisioning parts breakdown, and configuration change notices.

One of the mathematical models used for obtaining spare parts requirements is the METRIC model and its many variations. This model includes variables, which the life cycle material manager must provide for the team. These variables are:

1. Redundancy factor.
2. MTBF.
3. Confidence level.
4. Base repair fraction.
5. Repair times.
6. Delay times.
7. Cost factors.

The output of a computer run should be reviewed by the managers to confirm the accuracy of the input data. More information on models is provided in another chapter.

7.3.10. Modification

Modification requirements may be raised by the life cycle material manager or other personnel. It is the responsibility of the manager to evaluate and approve all proposed modifications. The cost of modifications to improve reliability or maintainability can be very high. For example, a change of an existing MTBF from 200 hours to 500 hours to improve availability, can be extremely costly (when considering redundancy). Some modifications, however, are not as costly. For example, a change in a locking device which allows faster removal of a printed circuit board to improve maintainability, can be a low cost modification. Once approval has been given to a modification, the items required for the modification and the documentation must be prepared. The supply manager must be informed and modification kits must be made available.

7.3.11. Repair and Overhaul

Repair and overhaul generally consists of maintenance which is difficult or impossible to carry out by a normal maintenance organization; the military refers to this as the third-line maintenance. The impact that repair and overhaul has on reliability and maintainability must be considered by the life cycle material manager, who must provide the technical content of the Statement of Work to the contract document. The manager is also responsible for reviewing and approving the repair and test schemes.

The supply manager is responsible for administering the repair and overhaul contract. The collection of failure data and replacement of parts information must be contracted for, and this information should be fed into the applicable maintenance information systems. The question of item availability for in-service use during the repair and overhaul period depends on the availability level required. That is, the item may be critical, urgent, routine, or pending depending on its availability. A priority system may be useful, to forecast repair requirements, especially when demands for serviceable items from stock cannot be met. In some cases, repair and overhaul results from the availability of new spares. Such cases of repair and overhaul must be technically evaluated by the manager. The reliability prediction by the contractor, which includes the failure pattern found, must be reviewed by the life cycle material manager.

7.3.12. Quality Assurance

Generally, quality control is a contractor's responsibility, and quality assurance is the responsibility of the customer. The quality assurance authority is responsible for establishing policies, procedures and standards and for implementing the quality assurance program. Both the quality control and quality assurance authorities sometimes function independently of the life cycle material manager, but in fact, should be under his control to assure that, in addition to reliability and maintainability, the approved policies, procedures, and standards on specific contracts for specific items are effectively implemented. Some of the activities of those authorities are:

1. Evaluating the adequacy of the contractor's procedures.
2. Reviewing the contractor's compliance to the procedures.
3. Verifying the characteristics of the products.
4. Assuring that the specification requirements are met.
5. Correcting deficiency.
6. Reviewing waivers and deviations.

All life cycle material managers should realize that the quality assurance tasks provide assurance that the material and the services procured from industry, and material repaired and overhauled, conforms to the technical requirements of the contract.

7.3.13. Material Storage Conditions

Material storage conditions is one of the most neglected subjects by managers. Distinct from the shelf life of a material is its material life, which is the time that an item can be anticipated to remain fully serviceable after it is put into use. The concept of material life is the basis for preventive maintenance.

Someone in the organization must be responsible for assigning shelf life, material life, and storage characteristics to any inventory item, and must ensure that items with limited life are managed as such. The assignment of special codes, including hazardous and dangerous materials, is essential. Codes can express time limits and characteristics, as well as storage requirements. Such codes should be part of the procurement procedure and should automatically be stated in each appropriate case. It is obvious that a material with a limited life has a major influence on reliability and maintainability.

The following two examples show the importance of proper storage. First, magnetrons stored on a high shelf were retrieved from the shelf when required, by a motor driven forklift. During the retrieval process, the magnetic field of the motor ruined the magnetrons. If a storage characteristic code had been assigned to the tubes, this could have been avoided. Second, rubber grommets stored in an uncontrolled environment for many long storage periods, were shipped to maintainers who found them to be unsatisfactory. Thus, new grommets had to be ordered and the large stock of remaining grommets had to be destroyed. A code would have prevented this additional maintenance time, delay, and cost.

7.3.14. Test Equipment

Life cycle material managers must get involved with two aspects of test equipment. One aspect is the provision of test equipment for every system or equipment introduced for the purpose of effective maintenance. The second aspect is the management and calibration of the system or the equipment.

Manufacturers and suppliers of new systems and equipment usually specify the type of test equipment required for the maintenance of the system or the equipment. Each manufacturer and supplier may have its own preference for certain test equipment which may not necessarily be the choice of the life cycle material manager. When choosing the appropriate test equipment, existing test equipment in the inventory must be taken into account, as well as the compatability of the test with other equipment especially the software (test equipment often have equipment, different operating test languages).

The reliability and maintainability of test equipment can play a vital role in the availability of systems in cases where only one particular test equipment is accessible to the maintainers, and this equipment is unreliable, requiring frequent repairs and recalibration. Redundant test equipment is costly, and some test stations can cost millions of dollars. Although much effort is being made to incorporate tests into the actual hardware, external test equipment remains a major life cycle cost requirement.

Most test equipment requires periodic inspection and calibration, and thus, management of a calibration program is essential. The life cycle material manager must identify when an item requires calibration. Such information can be stored in a management information system, especially if large number of test equipments are involved.

7.3.15. Contract Warranties

A warranty is a guarantee normally written into a contract, which furnishes the buyer with some assurance about the performance he should expect from a product. For example, the contractor might guarantee that a system will have a demonstated MTBF after a specified test period. It is common that contractors will attempt to negotiate the details of such a warranty with the customer. The details for such warranties can be very complex and may include some of the following:

1. Number of test samples.
2. Definition of a failure.
3. Test conditions.
4. Modifications.
5. Funding.
6. Level of confidence.
7. Reliability growth data.
8. Failure symptoms.
9. Record of time-related functions.
10. Noncompliance clauses.
11. Decision authorities.

The life cycle material manager has a major responsibility in areas such as

1. Monitoring the effectiveness of hardware or software covered by the warranty.

2. Activating corrective actions if required.
3. Negotiating corrective actions with the contractor, on engineering and technical problems.
4. Discussing financial aspects.
5. Evaluating all causes of failures.
6. Preparing progress reports.
7. Ensuring that the warranty clauses are in the interest of the buyer.
8. Making changes to contracts if required.
9. Recommending a particular type of warranty to suit special project needs.
10. Developing warranty clauses.

The provisions selected for a particular warranty contract by the manager must be realistic and feasible and not redundant, and must ensure that the resources are available to monitor the warranty. Maintainers must be made aware if a system or an equipment is under a warranty. Strict provisions must be developed to avoid cost increases.

7.3.16. Specifications and Standards

Reliability and maintainability specifications and standards must be available to all managers, so that their overall content and aim will be familir enough to ensure their appropriate applications. Hence, a list of all applicable and useful documents of this nature should be at hand and frequently updatd. Such documents are often revised by national, international, industrial, and government organizations. The listing of outdated documents is not only confusing but undermines standardization attempts. Technical tailoring of technical documents is sometimes done but any deviations from the documents must be well planned and substantiated.

7.3.17. Technical Data Package

Tenders of contracts require access to a large number of data. The life cycle material manager is responsible for retrieving approved data when required. The data package may include the following:

1. Specifications.
2. Standards.
3. Drawings.
4. Quality assurance provisions.
5. Reliability and maintainability documents.
6. Packaging data.
7. Special lists.
8. Computer tapes or discs.

The completeness of the package and all restrictions under which data is supplied must be verified by the manager.

7.3.18. Material Authorization

Accountable material issued in specified quantities must have some type of authorization. Often the types and quantities of material approved for issue to each user are fixed. A record of these entitlements must be kept, and if changes are needed due to shifting operational requirements, these requirements must be reflected in the authorized issue of material. Any authorization requirement presented to the manager must be analysed and evaluated in terms of need and cost, before final authorization is given.

7.3.19. Technical Orders

Technical orders are publications that deal with the operation and the maintenance of systems or equipments. Although the life cycle material manager is responsible for the approval and acceptance of such publications, he may be unable to write all the necessary technical documents without the help of the user or a contractor, who are both more familiar with all the technical details. Reliability and maintainability are greatly influenced by the operation and the maintenance instructions which have been developed.

It is common for a complex system to undergo certain modifications over the years. Consequently, progressive deterioration in the authenticity of the technical orders takes place. The manager must recognize this deterioration and take appropriate action to rewrite the documents. For example, a requirement existed, which called for the removal of an engine from a vehicle for maintenance purposes, after a certain number of kilometers. The engine was not to be opened before the temperature had dropped to a certain, specified level. During maintenance, it was found that the temperature specified was too high, which often caused the engine cover to crack, thus resulting in an engine failure. The temperature was then lowered by a number of degrees (Celsius), but this change was not noted in the technical orders. Other maintenance facilities, unaware of this change in temperature, continued to operate under the stated conditions, and thus experienced many engine failures.

7.3.20. Contracts

The process leading to the issuance of a contract that incorporates reliability and maintainability begins with the identification of the requirements necessary for completion of the project. These requirements may be in the form of a Statement of Work. The contract or contract demand establishes the procedures that will be followed for the control of the procurement activity. The Statement of Work specifies what is required, and the contract or contract demand states how the requirement is to be satisfied. Standard and special clauses combined with the Statement of Work make up the contract. The life cycle material manager will be confronted with some of the following in a contract which includes reliability and maintainability requirements:

1. Suitable methods of procurement.
2. Reliability and maintainability requirements.
3. Price for contracting goods services.
4. Contract time periods and identification of milestones.
5. Administrative aspects of contract.
6. Contract negotiations.
7. Review of procurement documents including the technical data package.
8. Proprietary rights.
9. Warranty considerations.
10. Contract clauses.

The last item has been a major problem area, and as explained in the Supply Manager Section, great care must be exercised to avoid that clauses introduced by people other than the engineering managers, are not overriding technical requirements, which downgrade or negate reliability or maintainability requirements. To write a contract in which there is no ambiguity and no misinterpretation possible, is not an easy task; but it is essential to avoid later disappointments.

7.3.21. Engineering Drawings

Engineering drawings are expected to contain the information needed to design and produce a reliable and maintainable item. The definition of an engineering drawing should be extended to include information relating to design, production, modification, testing, and material identification. This information should help the life cycle material manager in his effort to obtain and manage his material. Certain engineering drawing practices should be established, and instruction should be prepared for the preparation and management of engineering drawings.

7.3.22. Specialty Training

New systems and equipments may require the training of personnel in operating theory and in maintenance techniques, not yet incorporated in the general training. Software maintenance and complex electronic systems require very special knowledge, some of which is only recently being developed. Reliability and maintainability, software engineering, life cycle costing, testability, and other subjects are being taught at various centers. The life cycle material manager is responsible for recognizing the need for specialty training which will enhance his systems maintenance skills. Today, high reliability soldering or microprocessor courses are an almost must for any maintenance technician.

7.3.23. Budget and Funding Requirements

Normally, a financial organization will call for estimates once a year, to establish the funding needs for the years ahead, in addition to the estimate for the next year. The life cycle material manager will become involved in the estimates which outline his activities. Substantiation of cost estimates must be prepared because estimate input is subject to review. Regardless of whether the funding requirements are major or minor, funding for reliability and maintainability is often cut to a minimum in favor of other expenditures. It is the responsibility of the life cycle material manager to ensure that adequate funding is made available for all reliability and maintainability tasks, especially for demonstration, testing, and analysis.

7.3.24. Summary

The various activities of the life cycle material manager have been discussed in terms of the tasks required to achieve a high degree of reliability and maintainability, within a given cost constraint, over the life cycle of a system or an equipment. As can be seen his responsibilities are extensive, but necessary to achieve his aim.

7.4. THE SUPPLY MANAGER

The supply manager fits into the life cycle at the Acquisition, the Operating, and the Maintenance and Disposal Phases. His management activity starts when a product is acquired and installed, and also when plans are being made for a product's support. The supply manager's function, in the context of his involvement in reliability and maintainability, is mainly dictated by the procedures, policies and, instructions developed by the project and life cycle managers. Engineering management documentation must be translated into supply management tasks, and respond to the requirements of the final user and maintainer. Decisions that have engineering implications must not be made by the supply alone. It must be recognized that certain decisions are jointly made between supply and engineering, and others are made by supply management only.

7.4.1. Supply Management Tasks

A number of supply management tasks can directly influence reliability and maintainability management. These tasks are

- Initial provisioning.
- Spares allocation.
- Inventory location.
- Economic order quantity.
- Storing, packaging, and shelf life.
- Transportation.
- Inventory control.
- Catalogue identification.
- Contract clauses.
- Supply policy.
- Disposal.

Each task will be discussed separately.

7.4.2. Initial Provisioning

Not long ago, initial provisioning was carried out by a number of technical persons, who normally came from the supply side, and who had been asked to provide the spares for an initial operating period of the equipment. The determination of the quantity of all repairable and nonrepairable spares to be purchased was carried out mainly on the basis of historical information, as well as the criterion; that is, if the part rattled when knocked, more spares would be required. Much of this guess work resulted in an ineffective spares program, which in turn, played a role in the availability of the new equipment. For example, when a new, not previously known part was to be spared, in this case an early transistor, the technical sparing group felt that this may be a high risk item which may fail often. After ten years of equipment use, 114 transistors were drawn from stock, out of a 100,000 initial stock. In addition, the remaining transistors, when tested after ten years of shelf life, showed a high percentage of failures due to leaks.

Clearly, management was forced to demand a more effective way of establishing initial spares provisioning concepts. The demands that were made were simple—only as many spares as are absolutely needed for the operation of an equipment are to be purchased. Moreover, the number of spares purchased are to be readily available and within a given budget for a specified initial period.

Initial sparing involves three parts: the pipeline spares for repairable items, long lead-time spares, and a certain period of supply of nonrepairable items to keep equipments in operation while the in-use spares support is being established. The factors that influence sparing are failure rates, repair times, utilization rates, repair-line repair fractions, transportation times, operating locations and spares cost, or the highest availability for a given spares cost; all of these factors are directly tied to reliability and maintainability.

If a high degree of equipment or system success is demanded, and if spares must be available in a given time and within certain cost constraints, then the simplistic approach previously used by supply personnel must now be replaced by more scientific procedures. These scientific procedures involve engineering and logistic analysis, and are based on models which have been, or are to be developed. Currently many such models exist, which relate the number of spares required to both the failure rate and the time, to obtain the probability of having a spare of a particular part, available when required. Associating the cost factor, the various stocking levels, and the locations add additional parameters to the models.

The analyst, together with the life cycle manager, must decide from a list of models, which model should be recommended to management for use in a particular application. If the supplier carries out the calculations for the customer, the terms and the assumptions of the purchase must be clear to him. The supply personnel now become the action group who will implement the recommendations made by engineering.

7.4.3. Spares Allocation

One may assume that a chain of radar stations all require the same number of spares for a particular period of operation. Many studies have indicated, however, that the same equipment, used in different locations, may require a different number of spares. There are a number of reasons why this may occur.

Not all radars are installed in the same environment; that is, some may be installed in air conditioned facilities and others are not. The operational cycle may be different for each radar. The skills of maintenance personnel may be different at each site, and the maintenance levels may differ. Not all may be experiencing the same elapsed time, and some have gone through modifications, or overhaul and others have not. These differences influence the number of spares required, and thus, they must be taken into account by the Life Cycle Manager when planning for spares for each radar.

The allocation of the most effective spare quantity for each location is a task that must be carried out. Allocation must therefore take into account the following:

- The field experienced reliability and maintainability of each unit.
- The location and environment of each unit.
- The period of operation of each unit, if different.
- The difference in operation of each unit, if applicable.
- The available manpower skills at each unit site.
- The confidence level of failure identification at the spares level.
- The state of each unit—modified, overhauled, and elapsed time.
- The required availability.
- Procurement leadtime and delays.

Adjustment to allocation may have to be made by management to accommodate changes in the system mission, its operational cycle and its design modifications. In every case the key is availability. The importance of failure identification to spares allocation must be stressed, because it is well-known that the so called "no fault found" symptom can play a major role in sparing. The supply manager must be made aware of all these factors.

7.4.4. Inventory Location

In addition to providing the needed spares, consideration must be given to the distribution of the spares and to their location. Spares are not always located in one single location. If an equipment has been installed across the country, and possibly in other countries as well, a single spares location, or holding depot, would be faced shipping spares to places where maintenance is carried out. In the military, where three lines of maintenance exist, each line is usually not in the same location. Although the transport of spares consumes time, spares must be available at each location. There is a tendency for maintainers to order spares for the "just in case need." The result is drawers full of spares, which are not fully accountable for in the stock inventory. Such a habit is not very conducive to good spares management.

The choice of a spares location must be based on the spares requirements, the level of maintenance which is to be carried out at each site, the accessibility of the site, and the distance from a spares location to the maintenance site. In some isolated equipment locations, to which transportation may be difficult, spares of certain types may, nevertheless, be required (especially if the failure identification cannot be made from a remote area). Some of the remotely located telephone equipment have excellent BITE. The failure identification is transmitted to a spares holding depot, where the maintainer is informed as to which spare is needed at the site of the failure. Such a system is much more efficient than locating certain spares at the site of the failure. Any spares depot requires facilities and housing, and often, large amounts of space. To reduce the cost of such depots, their number should be limited without influencing their availability.

7.4.5. Economic Order Quantity

Any supply manager is interested in cutting the cost of spares. Thus, one of his inventory considerations might be to the concept of Economic Order Quantity (EOQ). As can be seen from Figure 7.1, there is a direct relationship between the inventory and the order cost for a given period of time. This relationship applies particularly to large quantities of common items rather than for high value items. Although this represents a simplified version of the subject, it can be used by a manager to assess his requirements in terms of the quantity of items that should be purchased, against the cost factors. The quantity can be calculated as follows:

$$Q = E \bigcirc Q = c \sqrt{\frac{2C_o D}{C_i}} \qquad (7.1)$$

where

C_o = The order cost
C_i = The inventory cost
D = The annual demand for an item
c = A priority factor

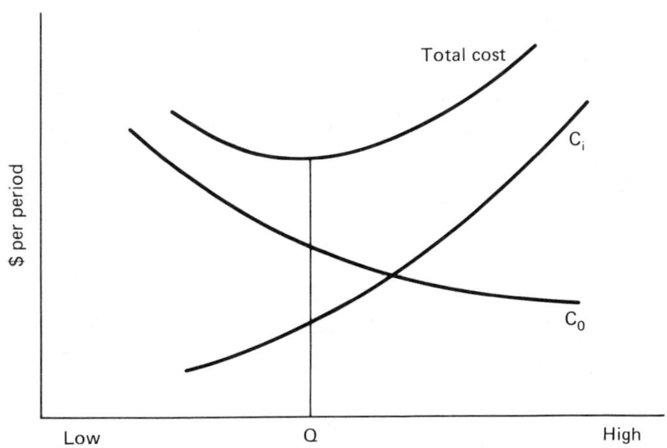

Figure 7.1. Relationship between inventory and order cost.

The priority factor is a scaling factor which expresses the importance of a spare to the operation.

7.4.6. Storing, Packaging, and Shelf Life

How stores are maintained influences the reliability of parts. Moreover, the ease at which stores can be withdrawn and located from a depot influences maintenance downtime. There normally are two types of stores, one that is general and one that is dedicated. The general store may apply to any and all requirements. The dedicated store or, "peculiar to" store, applies to a specific requirement such as a single quarter ton truck or a navigational aid equipment. The complexity of recent electronic equipment has increased the use of peculiar parts. The complexity of the equipment is due to the suppliers' attempts to capture the market by ignoring the standards of the design teams, and the demand for increased functional complexity, which requires peculiar items. The question of standardization will be addressed under Catalogue Identification.

Before any storing is undertaken three tasks must be carried out: supply management must insist on a stock number for each item, decide the packaging required for items, and assign a code to all shelf life sensitive items. The identification of proper packaging influences reliability. Not only is packaging important for storing but it is also important for shipping. The quantity packaged can influence reliability; that is, large quantity packaging is economical except when shipped to the maintainer who requires just a few parts from the package, and leaves the rest of the parts stored without control. In addition, certain types of packaging is unsuitable for some parts. For example, rubber material should not be packed up in high acid content packaging material. Moreover, the majority of ordinary papers are unsuitable as packaging material. Some semiconductors are sensitive to static electricity and packaging must therefore be selected to reduce this effect. Other material may be flammable or corrosive. Thus, special warning labels for each package must be made. A classic example of ignoring proper packaging was the storing of magnetrons in boxes identified only with a stock number. Large and heavy metallic fork lifts with electric motors were used to lift the magnetron from one of the shelves. The result—a failed magnetron. Such incidents are seldom noted by the supply side but when such items hit maintenance, the blame must rest with the life cycle manager, who neglected to instruct the supply manager in how to deal with such items.

Although shelf life sensitivity was recognized many years ago, little has been done to rectify this problem. Not only must a shelf life be assigned to such items but also a shelf life code, which expresses the anticipated shelf life and indicates if such items require special handling or packaging. The available shelf life data for material are limited, although a number of countries have some listing for United States, United Kingdom, and Canada military materials. Dormant reliability data may be found in a few publications. The analysis of dormant reliability factors for materials stored usually under uncontrolled temperature and humidity conditions is essential for the establishment of shelf life information based on scientific methods rather than on good guesses. At every initial provisioning conference the shelf life of required parts must be considered. Deterioration of material on a shelf can be very costly and can also downgrade reliability. For example, it is senseless to store rubber material for extended periods on very high shelving that experiences high temperatures in a depot during the summer months. Also, it is inadvisable to store servomotors for long periods without rotating them, as rotation ensures that lubrication does not flow to the bottom of bearings, thus reducing life considerably. It has been suggested by some writers that a retest/storage cycle could result in better storage life compared to simple shelf storage. The retest/storage cycle involves idea storage, then periodic tests (with possible maintenance), back to storage, and then installation. The storage reliability can be calculated using

$$(1 - a)\,\lambda_s\, t_2 + \lambda_s\, t_3 \tag{7.2}$$

where a is the portion of failure detected by the test

- t_2 is the non-operating time prior to installation
- t_3 is the non-operating time between the last periodic test and installation.
- λ_s is the storage failure rate.

7.4.7. Transportation

The influence of transportation on supply management and in turn, on system availability must be considered. The two factors that have a major management impact are the transportation times and the transportation costs. Transportation from sources of supply to the different levels and lines of maintenance influences the logistic support. The time factors must be added to the total maintenance time but it is advisable to record such time factors separately to avoid confusion with the actual maintenance time. The supply manager must take into account the transportation time when ordering items for the inventory or when shipping items from the inventory.

7.4.8. Inventory Control

Without an inventory control system there is little control over the reliability of ingoing or outgoing items. Here again, the concept that supply management must act on advice and instructions from the life cycle manager must be recognized. If the life cycle manager requires parts that must meet certain reliability levels throughout the life of his equipment, he must provide instructions on how to assure such reliability. Too often, supply management decides how to order parts and how to supply parts required for maintenance. In most cases, supply managements decision based on the idea that the cheapest part will do as long as it is "equivalent." The question of contract clauses will be discussed in another section.

In order to effectively control the inventory, the supply policy must be system/equipment oriented rather than commodity oriented. That is, the inventory must be tied to the system or the equipment rather than to a generic commodity. The commodity related inventory cannot identify what equipment an item belongs to, it cannot separate common users items used for one or several types of equipment, and it cannot itemize what the life cycle for a particular equipment is in terms of items used from the inventory.

There are many automated inventory control systems that use various techniques for item identification, but these are not discussed in this book which deals with the reliability management aspects only. Two identification problems have a direct bearing on reliability. One has already been mentioned, namely the problem of relating items to systems and equipments. A coding system should exist that makes this relationship possible; that is, an equipment applicability code (that uses alpha-numeric digits) should relate each inventory item to its actual use. The question is, How far down in the equipment breakdown should one assign such codes? The more detailed the breakdown, the larger the number of digits required. Five digits may be sufficient for a breakdown to the module level. This coding is essential since not all items, even if identical require the same reliability, and thus, purchasing all items to the highest required reliability would be very costly.

One particular supply datum, the quantity used, is very valuable to the analysis of the field reliability of inventory parts, and may be one of the few factors available for calculation of failure rates. Many equipments do not have elapsed time indicators and the time factor must therefore be established by using the operational equipment times and the mission times, or the usage factor. (Unfortunately, the usage factor may be misleading in that not all parts used were parts that failed.)

One other aspect of inventory control that affects reliability is the principle by which an item can be selected from stock out of stores. When supply purchases items, it is not always possible to obtain exactly the same item, due to slight manufacturing changes or modifications, made to accommodate certain specifications. Parts with certain electrical values, may have to be changed to meet small changes in electrical values. For example, a 55 Ohm resistor can now be purchased only as a 56 Ohm resistor, which is a standard value. In this case, the original and the newly purchased item will be binned separately and usually given a new stock number. A direct replacement, however, is possible, and inventory control must recognize this. There may be three or four parts that can be used for the same purpose and with the same reliability, but each with slight changes.

A hierarchial structure is needed to assure that the sequence of withdrawal preference from stock is followed. This structure must be set up by the life cycle manager, who is fully familiar with an item and its characteristics, including its reliability. One danger is that the life cycle manager will withdraw the latest and best part first, rather than search for any other possible stocks. This results in older parts remaining on the shelf sometimes forever. A simple preference stock code is all that is needed for the identification.

7.4.9. Catalogue Identification

An important step that supply management must perform is the identification of all items entering the system. From a reliability standpoint, the correct identification of any items, and specifically, those items that have specified reliability requirements, is vital. The military is making use of a codification system, which is a uniform and common system for identification, classification, and stock numbering of stock items.

The identification scope adopted is for all items of supply for which there is a repetitive, recurring, or continuing requirement. These include:

1. Items that warrant central item management, reporting, or stock control for supply, maintenance, or procurement.
2. Items procured and stored for operational use, insurance, or reserve.
3. Items that will prove of value to one or more material logistics operations, including local purchase, locally manufactured, and maintenance and overhaul items repeatedly procured, stored, distributed, or issued.

The objective is to increase the effectiveness of the equipments or systems by providing each item with a single item name, a single uniform classification, a single uniform identification, and a single uniform stock number. Item names should use approved names and definitions, rather than manufacturers' names, which cannot be related to approved names.

Item identification should consist of minimum data which adequately describe the characteristics that give an item its technical uniqueness and differentiate it from all other items. Hence, each item should have only one item identification. Two methods of identification are available, the *descriptive method* and the *reference method.* The descriptive method uses uniform names and definitions, which reflect the technical or performance characteristics of an item. These names and definitions can be coded for computer files to provide automated screening by key characteristics and descriptive outputs. (Drawings should be developed for items that cannot be presented directly in words.) The reference method is an indirect process of identification, whereby item name is supported by reference data only. Some existing identification systems use a thirteen digit number, and do not include much data on the reliability, shelf life, and applicability, which are needed for procurement.

Catalogues that identify the item by its number only are useless to the user who is not familiar with the number that he wants to obtain from stock; that is the code seldom gives any clue as to what the item is in technical terms. The reference of a code stock number to a brief description, or an identification number plus an occasional illustration, will give the user a good idea of what the term is. Moreover, reference to a specification will narrow the information gap even further. If the item ordered is not in stock, and a replacement must be from the existing stock, identifying and locating the item in the catalogue may become a lengthy process. Automation of the catalogue such that the required technical data leads one item to another of the same or similar technical characteristics, may not be available, at least not to the extent that the user would like to see. Most larger, automated catalogues have some sorting capability, though hardly enough to satisfy the life cycle manager. The selection of an alternate item, having the same or similar technical and reliability characteristics, is important to avoid degrading equipment reliability.

The visual display of items, in sequence of their technical and reliability characteristics, allows the identification of items no longer required. Deletion from stock of such items saves not only space and auditing time, but also recoverable material, that if sold, can provide a major economic boost. Deletion of unnecessary stock also ensures that items are not used in equipment for which they were not intended.

7.4.10. Contract Clauses

A contract is written by the customer, for items that must be purchased. In addition to the technical description of the ordered item, a number of standard clauses are included, which in many cases, are legal clauses common to all contracts. Such clauses are added by the supply organization after the technical information has been completed. In practically every case, the technical manager fills out the portion of the contract pertaining to him and never sees the final contract (with all the added clauses) until it has gone out for a bid. Unfortunately, the added clauses may override, knowingly or unknowingly, the technical requirements; however, this judgment can be made only by the life cycle manager (the technical authority), and not the supply manager. A large number of incidents have occurred where the contract clauses added by supply, (the purchasing authority) have been detrimental to reliability. Therefore, regardless of the time factor involved, the technical manager should review the contract before, and not after, it goes on the street.

To give some examples a few such cases are mentioned. A clause that is often added by supply management to a contract is the "or Equivalent" clause. From a supply management viewpoint this makes sense because the possibility of buying a lower priced item than the original one, which can do the same function, is attractive. (This is especially true when buying nonstandard items on which a larger number of suppliers may bid). Recently, a large order was placed for fuses. Over one million were ordered. These were glass enclosed, nickle plated fuses. The "or equivalent" resulted in one company having in stock the same kind of fuse in very large quantities, but silver plated. Because the offered price was about half that of the type ordered (the company wanted to get rid of them fast) the silver plated equivalents were purchased by life cycle managers for Navy destroyers. (The rating, size, and enclosure was identical.) Within a few days all of the fuses had corroded due to the salt atmosphere. Another well-known added clause is "Certificate of Compliance Acceptable." This clause allows any quality or reliability test results to be waived; as long as a legal authority issues such a certificate, the item must be acceptable. Certificates of this type can be purchased from many legal authorities who have no idea of the technical requirements. High value tubes, costing $30,000 each or more, have been bought with such compliance certificates. The companies that sold these tubes to the government some years ago, used old tubes, rejuvenated them, and certified that these tubes complied with the technical requirements. There was no difference between these old tubes and the original tube and initially, when installed in the radar they performed well. But the operating life of these tubes was considerably shorter and varied over a wide time range. Thus, radar reliability was down, maintainability was up, and the life cycle cost increased. One more example of the added clause is the "Incoming inspection will verify quantities delivered against contract" clause. This clause allows for acceptance of items by count only. It does not require technical procedures, and does not provide for technical noncompliance of items (no time period specified for return of items not meeting the technical characteristics). Of 5000 standard ICs which were stored for four months, almost 80% were found to be contaminated when installed in equipment. The ICs had been accepted and the company refused to replace them since no such provision was made in the contract.

When buying items, careful consideration must be given to the development of suitable contract clauses that ensure quality and reliability. The development of such clauses must be by the life cycle manager or the user.

Protective clauses should be developed, which allow replacement of items not meeting the purchase requirement. Guarantee clauses, which oblige the supplier to meet certain important needs, should form part of a list of available clauses. The question of reliability warranties is discussed in another chapter, along with the importance of contract clauses developed by management, which assure no decrease in reliability of equipments using items purchased for maintenance. Also discussed is the need to monitor contracts prior to their release to avoid addition of legal clauses which override technical clauses.

7.4.11. Supply Policy

The development of a supply policy, which can meet the reliability requirements, must be a joint management project of the supply and life cycle managers. A simple supply system for a TV set supplier, however, most likely has no special policy for the distribution and supply of repair parts. The inventory reflects the demand for parts by the repair houses, and the parts being held in stock are the lowest priced parts normally obtainable. It is an entirely different policy that governs equipments or systems supply; that is, a certain reliability requirements as specified by the life cycle manager, must be incorporated into the supply policy.

Most of the areas that impact on inventory, and that must be reliability oriented, have been discussed in the previous sections. The supply policy must incorporate general statements which cover initial provisioning for specified, reliable parts; the storing and packaging of parts; the identification of shelf-sensitive items; the description of items with their reliability codes in the catalogue; the ordering of parts with the appropriate contract clauses; and interfacing with the life cycle managers on technical decisions. Naturally, the supply policy will deal with other matters as well, although these are not discussed in the context of reliability and maintainability management tasks.

7.4.12. Disposal

The disposal of unnecessary items requires some consideration by the supply manager and life cycle managers in the area of reliability. The disposal of an equipment which cannot be supported, or has become obsolete, may require destruction or sale to other parties (e.g. the sale of wire to recover a raw material, such as copper). When the sale is to another party who intends to continue to use the item, the new party may want to obtain all the accumulated data about the item from the previous owner, in order to better plan its operation and maintenance. In such cases, the reliability and maintainability data are valuable. The data must come from the life cycle manager, who transfers them to the Supply Manager, who in turn transfers them to the new customer. With the transfer of an equipment may go all the required spares, modification kits, special test set-ups, and all the documentation. If applicability codes exist for all items used in the equipment identification is easy. If such codes do not exist, and instead, much common user items are included, a statistical review must be made of the usage rates so that stock can be withdrawn to meet a certain operational period. Arrangements such as these vary in extent from one to the next case and it is a management task to determine the most advantageous transfer arrangement.

7.4.13. Summary

The dependence of the supply manager on requirements generated by the life cycle manager has been demonstrated. Technical decisions must come from the technical managers and not the supply managers to avoid downgrading reliability or maintainability. The various areas with impact on supply have been discussed.

7.5. EXERCISES

1. When considering a new equipment during the Concept and Definition phase, the project manager must be aware of certain constraints which will affect the logistic support once the equipment is fielded. Name a number of these constraints and describe how they affect the logistic support.
2. The life cycle material manager comes into action just prior to the In-Use phase, with a fairly large number of tasks to perform. List some of the tasks that directly influence reliability and maintainability.
3. No item will last forever, that is, when items are stored certain deterioration takes place. What kind of quantitative evaluation should be made of items, and how could these quantities be expressed to guide management in storing and handling them?
4. A warranty is a guarantee normally written into a contract. What kind of reliability or maintainability guarantees could be written into a contract?

7.6. REFERENCES

1. Naval Publications and Forms Center. MIL-STD-1388, *Logistic Support Analysis*. Philadelphia, PA.
2. Naval Publications and Forms Center. MIL-STD-1369, *Integrated Logistic Support Program Requirements*. Philadelphia, PA.
3. Naval Publications and Forms Center. DOD-STD-480, *Configuration Management*. Philadelphia, PA.

CHAPTER 8

Management of Reliability and Maintainability Documents

8.1. INTRODUCTION

The management of documents is critical to the achievement of effective reliability and maintainability. As the design proceeds, it becomes defined by an appropriate set of documents. Design analysis proceeds by reviewing the data which flow from the documents. The principles of reliability and maintainability document management include the establishment of baselines, control changes to these baselines, record and track status, and audit.

8.2. DOCUMENT PROLIFERATION

The number of documents required for any hardware/software may be very extensive, especially if the hardware is complex and goes through many different organizations. These documents have been developed and updated by individuals who required certain information for their operation, and thus, little thought has been given to ensuring that all documents have the same baseline to achieve effective reliability and maintainability. This kind of procedure often leads to uncoordinated and unnecessary documents. Conflicting information may result downgrading the effectiveness of the hardware. A large amount of duplication increases the document volume and, in many cases, certain documents are not even called upon for very extended periods. A good example is the split of inventory documents which is often encountered. Three or more documents may have been developed that each take a space of more than hundred feet of shelf, or very extensive computer file. One document may contain all the stock class in sequence numbers; another the item identification in detail; and, still another, the physical description and sketches. The documents may contain all the stock class numbers in a numerical sequence, or a sequence arranged by technical considerations. A document may also identify the items in technical detail, or list the items in tabular form. (Note: these four different documents may be used only occasionally by managers.) For certain countries, in which more than one official language is spoken, the number of documents may be double or triple. Document proliferation is very costly and, in some cases, may outweigh the hardware cost. Therefore, the number of documents used must be a major management concern.

8.3. DOCUMENT TYPES

It is not this book's intention to list all the types of documents required and for which management must give an account, but rather to outline a few major types which are often encountered. Such documents are as follows:

1. International—standards, publications, or specifications.
2. National—may be standards, publications, or specifications. (Documents 1 and 2 may originate from government or industry sources and may be directed towards one or the other, or to both parties.)
3. Policy—cover overall agreements and requirements, and set the stage for lower echelon documents.
4. Procedures—explain the steps that must be taken for a particular application and the kind of approach which should be used. Most procedural documents are general and do not go into technical detail.
5. Instructions—provide detailed instructions for actions and tasks. These may be maintenance instructions for the maintenance of hardware or the instruction for preparing a computer file.
6. Plans—outline a plan for action and the steps required by individual managers.
7. Reports—deal with specific or general subjects but, for the purpose of this discussion, influence reliability and maintainability.
8. Lists—may be forms with information (in most cases technical data).
9. Drawings—required to reflect the hardware in details.
10. Contract—required for any contract.

The technical documents that must be considered may cover some of the following technical areas:

1. Installation instructions.
2. Test data and record forms.
3. Maintenance procedures.
4. Operating instructions.
5. Repair and overhaul manuals.
6. Software programs.
7. System and equipment manuals.
8. Test equipment instructions.
9. Technical standards and specifications.

Since all these documents may confront management, it is obvious that some sort of recognition must be given to effectively manage them.

8.4. DOCUMENT MANAGEMENT PLAN

Although all hardware may not need the extensive documentation that will be suggested in the following sections, practically all hardware will require some sort of documentation. The various subjects covered may apply to small, large, simple, and complex hardware, where reliability and maintainability are a prerequisite. Therefore, the content of a formal or informal management plan must be chosen carefully. The plan must be prepared by, or for, a top-level manager in order to allow for proper coordination of the total document package to control track and audit the required documents.

8.5. REQUIRED DOCUMENTS

As an introduction, the plan should identify all the documents to be produced or required, the organizations to which the plan applies, and the time period in the life cycle for which the document has been written. A complete listing of all documents by title, number, date, and publishing organization is needed.

8.6. ORGANIZATION

To achieve effective reliability and maintainability, the plan must describe the organizations that will develop, use, or select the required documents. Measures must be taken to ensure that the staff involved in documentation has the necessary background, and management control is at the correct level. Moreover, the organization must have the necessary resources to perform the required tasks.

The documentation manager requires a number of supporting personnel who can deal with some of the following management tasks:

- Preparation.
- Support.
- Data services.
- Drafting and graphic services.
- Format design.
- Editing.
- Production and reproduction.
- Control.
- Status identification.
- Audit and reviews.

Data services must have the required data in order to prepare, control, review, and report the status of documentation. Format design is responsible for effectively formatting any documents, (including drawings). The organization responsible for documentation management must work together with the life cycle managers, who have particular requirements for documentation.

8.7. DOCUMENTATION CONTROL

Reliability and maintainability information, which may be part of any documentation, changes over the life cycle. Therefore, it is important that management controls these changes and identifies them to ensure that the hardware matches the software. To control such action, the military has issued two standards—MIL-STD-480 and 481—both dealing with Configuration Control-Engineering Changes, Deviations, and Waivers. Various types of changes are classified for review and approval by management.

The following example illustrates how document management planning functions. A system specification is prepared after a system has passed through the concept and definition phase, and is in the acquisition phase. Thus, the reliability and maintainability requirements originally specified, must now be met in the acquisition phase. Production samples may show different reliability or maintainability values from analysis or tests and these changes must be reflected in the original specification, because they may affect logistic support requirements. Therefore, documentation control must ensure that such required changes are entered, as necessary, in the documents. (This is a management responsibility.) The documentation cycle is shown in Fig. 8.1.

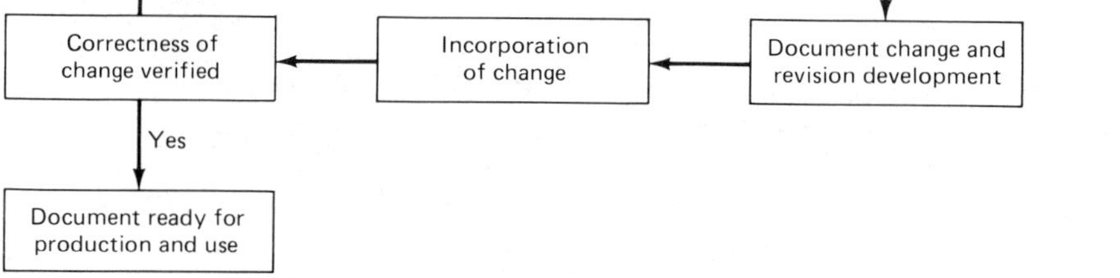

Figure 8.1. The document change cycle.

8.8. DOCUMENTATION STATUS ACCOUNTING

Status accounting involves recording and reporting the document information, so that the system is managed more effectively. Data must be submitted to management for collation and preparation. Management must also accept input from others when collating such data, and must be responsible for distributing the documents.

There is a need to identify the reliability and maintainability data required for status accounting, and the method of how it should be recorded. It should identify the distribution and frequency of such status accounting reports and the format.

8.9. DOCUMENTATION AUDIT AND REVIEW

Documentation audit and review must be defined in terms of what is to be performed at specified points in the documentation life cycle. The items to be covered at such audit and review must be identified, and the procedurs to be used for the identication and resolution of any problems arising from an audit or review of documents containing reliability and maintainability data, must be stated.

8.10. CONFIGURATION MANAGEMENT

What has been described in the previous paragraphs is basically the concept of configuration management. The definition of configuration management as it applies specifically to documents is as follows: technical and administrative direction and surveillance to identify and document the functional and physical characteristics of a configured item, to control changes in documents of these characteristics, and to record and report changes set forth in technical documents.

Configuration management normally starts with the specifying of a base line. The engineering documentation describes the characteristics of hardware or software, and all their elements, as they exist at a given time in the life cycle. These engineering details will range all the way down to the parts level with their part numbers. The identification of the configuration reflects the technical document content, and thus, any changes in the configuration must be formally established in the documentation. The evaluation, the coordination, and the approval of any change to the engineereing documents is called a *change control* in configuration management. In addition to the engineering documents, a maintenance logbook may exist, which among other things, is used for logging any modification. Modifications may lead to new part numbers or to a reconfiguration. Such reidentification due to modification must be embodied in the engineering documents as part of the configuration control and management. Design changes, deviations from a standard, or any waivers which apply to an equipment or a system built to a specification, must be controlled.

A well-managed configuration management plan should result in the following features:

1. Identification of all characteristics of a configured item. (The control, accounting, and audit will be simple.)
2. A minimum amount of documentation.
3. A relationship between various managers that will become easier to coordinate.
4. Improved methods for evaluating and dealing with any required hardware/software changes, which will reduce problems during the life cycle.
5. Identification of all required configuration changes. (Changes are better identified and accounted for.)
6. Goals to follow using simple procedures.
7. A completeness in definitions and change descriptions.
8. Greater management confidence due to more accurate reliability, maintainability, and logistic support documentation.

To assist management in the control of systems and equipment, codes may be assigned for each type of system or equipment, such as a code for the identification of an aircraft, a vehicle, a radar, a frequency generator, or a binary counter module. Assignment of such codes to items is a configuration management task, which helps control and manage items that have changes in their characteristics.

8.11. APPLICABLE MANAGEMENT DOCUMENTS

There are a number of specific documents that can be of some assistance to management. No specific list is given here, but the subjects covered in available documents and found in directives, specifications, standards, instructions, requirement documents, and books, are briefly outlined below:

1. Reliability and maintainability management.
2. Management of contractor date and reports.
3. System engineering management procedures.
4. Data management.
5. Management of data on contracts.
6. Data acquisition management.
7. Data prices and cost analysis.
8. Procurement data packages.

9. Data quality.
10. Engineering management.
11. Report, reliability, and maintainability data.
12. Engineering organization and management.
13. Management and office information systems.
14. Reliability and maintainability acquisition.

8.12. TAILORING OF DOCUMENTS

Recent years have shown that data requirements have become more detailed and complex. Previously, only the MTBF and MTTR values were stated for design, demonstration and testing. Reliability and maintainability efforts have been expanded over the years, and some of these efforts have led to standardization. This kind of approach is very rigid and for some time it appeared as if all procurement data requirements looked the same in every case. But projects differed in concept and performance requirements, and each project had different individual constraints to deal with. There was no room for any deviation from the rigid standard imposed by existing documentation. This rigid approach can be overcome by:

1. Encouraging engineers and management to show more flexibility in the development of reliability and maintainability data for projects.
2. Tailoring reliability and maintainability data to meet the specific requirements of the planned system.
3. Varying reliability and maintainability data from project to project and recognize differences between projects.
4. Maintaining a level of standardization while recognizing the need for flexibility.
5. Consider the effect of cost when pursuing alternate data requirements for a project.

Tailoring is a management tool that permits the stating of reliability and maintainability data requirements, so that specific project requirements are met. Reliability and maintainability requirements can thus be well-planned before a system is delivered for field use.[4] Although each reliability and maintainability data package may specify similar contractual requirements, each project will be somewhat different and often unique in its approach. Flexibility must be based on cost effectiveness and keeping some level of standardization in mind.

8.13. DOCUMENT UPDATE

It is not very often that management makes inquiries to determine if a document, to be used, is the latest version with the appropriate revisions and amendments; that is, it is assumed that the latest document is called up on to governs. Besides, few managers, if any, ever have immediate access to all the latest issued documents required by an organization. Documents issued by government agencies must usually be purchased after they are released, and changes in internal documents often go unnoticed. The number of documents that might be needed for a particular project can be extensive. Moreover, each document may call up on other documents, thus extending the list even further. In any contract, it is very important to clarify which document governs, and what parts of it do or do not apply.

Documents such as specifications and standards are continuously updated through revisions and amendments. A document with a revision may carry the same number as the previous issue, but the title, as well as the contents, may have changed. (This does not happen often but such cases do exist.) Changes to documents can be minor or major in nature, but most revisions are made to make major changes rather than a few minor alterations. Technical changes may not always be acceptable for contractual purposes. Sometimes, the older document version is preferable to the revised one, even if it means nonacceptance of the latest standard. If such is the case, this must be clearly stated to avoid any misunderstanding and possible contractual difficulty.

It is therefore important that management becomes familiar with all updated documents as soon as they appear (especially those that are required for a particular project). A review of such documents should be made prior to listing them for contractual purposes. If necessary, documents may have to be tailored, as explained in section 8.12. Tailoring is a strict management function, which should not be taken lightly (especially in view of the many liable suits being fought over the subject of "latest document status").

8.14. SUMMARY

Documentation management is critical to the achievement of effective reliability and maintainability. Some of the tasks that must be performed to achieve effective reliability and maintainability have been identified. Careful planning can reduce the cost of documentation.

8.15. EXERCISES

1. What kind of functions must management carry out to deal with reliability and maintainability documents?
2. Prepare a list of the kind of documents that deal with reliability and maintainability and which management should be aware of.

3. What kind of steps must be taken to change a document? Prepare a simple flow chart that outlines these steps.
4. Idenify the other people, aside from the life cycle material manager, who must provide input when preparing a document dealing with reliability and maintainability of equipment or a system.
5. Define configuration management in terms of management and documents? Explain the steps that are part of configuration management.

CHAPTER 9

Management Tools for Reliability Managers

9.1. INTRODUCTION

Effectively managing a reliability assurance program, is a challenging task to reliability managers. The factors responsible for this challenge are (1) the emphasis on cost effectiveness; (2) the size and complexity of systems; (3) the stringent reliability requirements in the design specification; (4) the competition and stringent deadlines associated with the completion of a project. Therefore, in order to perform the reliability management function effectively, reliability managers must use the modern tools of management, as well as useful reliability assurance procedures. This chapter presents selective reliability assurance procedures and selective management tools. The chapter is divided into four parts, which are concerned with procedures for reliability assurance, configuration management, value engineering, and network analysis technique. The first part consists of various useful procedures, concepts, and other information strictly concerned with reliability assurance. The remaining three parts are concerned with the general management tools that reliability managers can make use of.

9.2. PROCEDURES FOR RELIABILITY ASSURANCE

This section presents selective procedures, concepts, and other information, which is useful for reliability assurance.

9.2.1. Designing for Reliability—A Six Step Approach

This approach consists of six major steps. The approach has been found to be successful in several cases.[1] The steps of the procedure are shown in Fig. 9.1. These steps are concerned with establishing stress requirements, developing preliminary design, estimating failure rate, identifying failure modes

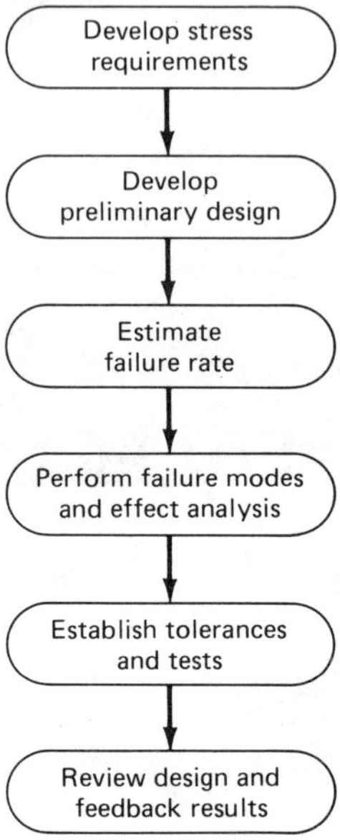

Figure 9.1. Designing for reliability approach.

and effects, establishing tolerances and tests, and reviewing design and feedback results. Most of these steps are self-explanatory, however, some of them are described below.

Develop Stress Requirements. Here, the designer is faced with determining the stresses to which the item under design will be subjected, in its operation mode. Usually the stress requirements are provided in the design specification, however, quite often they are based on the total system assumed condition of operation.

Examples of stress requirements are shock, vibration, temperature, moisture, and humidity.

Estimating Failure Rate of the Item Under Design. After accomplishing the first two steps (i.e. developing stress requirements and preliminary design), the next logical step is to make a preliminary estimate of the failure rate of the item under design. This can be accomplished with the aid of any one of the following three techniques:

1. Experimental test technique.
2. Random-failure technique.
3. Stress-strength technique.

The first technique is concerned with actual experimental testing of the prototype item. The benefit of this approach is that it ensures good reliability estimates of failure rates of the item in question. The disadvantage of this technique, however, is that it is both expensive and time consuming. The second technique is based on the fact that the failure rates of parts are constant; therefore, the overall failure rate is given by the sum of the failure rates of the individual elements. Thus, one must estimate the failure rate of the individual elements. The third technique is concerned with determining the stress-strength relationship during the useful life of the item. In order to determine the probability of failure of the item under consideration, the associated stress and strength probability density functions must be estimated.

Failure Modes and Effects Identification. In this step, the objective is to identify all failure modes of the item and their effects. The technique of identifying failure modes and their effects is commonly known as the *failure modes* and *effect analysis* (FMEA) *approach*. Proper application of this approach helps to reduce the probability of failure of the item under design.

9.2.2. Useful Concepts to Improve Equipment Reliability During the Design Phase

The following methods are used to improve reliability:[2,3]

1. Fail safe.
2. Redundancy.
3. Safety margin.
4. Safe life.
5. Durability.
6. Simplicity.
7. Derating.
8. Specifying standard items.
9. Limited life components.

The *fail safe* method is concerned with designing an equipment in such a way that the failure of a single element does not cause the host equipment to fail. *Redundancy* is one of the most useful methods to improve the reliability of the equipment. With this method, more than one item is used in the equipment design to perform the same function. Two commonly used redundant configurations are known as the parallel and standby. The *margin of safety* method is another technique used to minimize failure. It is defined as follows:[2]

$$M_s = \frac{S_m - S_{sm}}{S_m} \tag{9.1}$$

where

M_s is the margin of safety
S_m is the minimum strength
S_{sm} is the maximum stress.

Sometimes components used in a product have a failure-free life that is greater than the life of its parent equipment. This type of approach is known as the *safe life*. The *durability* method is concerned with the selection of material and performing design details in such a way that the resulting system is resistant to degradation from factors such as fatigue, erosion, wear, and corrosion. The *simplicity* method uses parts and assembly simplification to improve reliability. In otherwords, the simplicity of the design will help to increase reliability. *Derating* is another method that helps to improve the reliability of an item. In this case, the maximum operating conditions such as temperature and power are derated below their specified values.

Specifying in the design, involves using standard parts to increase reliability. Another method practiced to increase reliability makes use of *limited life components*. In this case, the parts designated as limited life items are replaced after a specified period.

9.2.3. Useful Guidelines for Designers and Management

This section lists a number of guidelines for designers.[4] Some of these guidelines are as follows:

1. Be aware that an increase in system complexity decreases its reliability.
2. System data collected from field should be analyzed with care.
3. It is wise to use a large number of simple parts, rather than a small number of complex parts in a system under design.
4. It is advisable to determine and control factors affecting reliability in the early stages of system design.
5. Determine and reinforce weakpoints in the equipment design.
6. To increase part reliability in the region of 99%, is time consuming, costly, and difficult in comparison to 90 or 95%.
7. Avoid making the designed item too inflexible; accept specific penalties to make it flexible.
8. Provide displays or indicators to signal impending failure.

9.2.4. Achieving Reliability Objectives in Government Contracts

To achieve reliability objectives in government contracts, attention must be directed toward the following items:[1]

1. Risk analysis and quality control.
2. Standardization and design simplification.
3. Motivating manpower on a continuous basis.
4. Periodic auditing of methods and procedures.
5. Proper utilization of experienced reliability manpower.
6. Making use of redundancy whenever it is necessary.
7. Conservatively choosing safety factors.
8. Top management involvement in support of reliability.
9. Proper utilization of specialized knowledge inside as well as outside of the department.
10. Collecting data on failure incidents.
11. Resisting design changes when reliability is considered to be satisfactory.
12. Revising failure rates whenever it is necessary.
13. Coordinating the organizational elements.
14. Screening and burn-in of components.
15. Testing to verify the required reliability.
16. Education.
17. Initiating and executing corrective measures.

9.3. CONFIGURATION MANAGEMENT

Hundreds of engineering changes are associated with the development of a complex system. Configuration management assures both the customer and the manufacturer that the resulting system hardware fully meets the contract specifications.

The United States armed forces played an important role in developing the configuration management discipline. In fact, as early as June, 1962, the United States Air Force produced a document entitled "Configuration Management During the Development and Acquisition Phases," AFSCM 375-1. Since then, several important documents have been produced on configuration management. Selective documents are presented in Table 9.1.

9.3.1. Terms and Definitions

This section presents selective terms and definitions associated with configuration management disciplines.[5] *Configuration Management:* the art of providing systematic configuration identification, control, and accounting to a system and its components. Configuration management is defined as the management of technical requirements which define the engineering system as well as changes thereto.[6]

Table 9.1. Selective configuration management documents.

Item No.	Document	Title	Produced By
1	AFSCM 375-1	Configuration Management During the Development and Acquisition Phases	United States Air Force
2	NPC-500-1 (Now NHB 804.2)	Apollo Configuration Management Manual	NASA
3	NAVMATINST 4130.1	Configuration Management	United States Navy
4	AMCR 11-26	Configuration Management	United States Army
5	DOD Instruction 5010.21	Configuration Management	United States Department of Defense
6	BUWEPSINST 5200.20	Weapon system Configuration Control Manual	United States Navy
7	DOD Directive 5010.19	Configuration Management	United States Department of Defense
8	OPNAVINST 4130.1	Configuration Management of Software in Surface ship Combat Systems	United States Navy
9	NAVELEXINST 5420.5A	NAVELEX control and review Boards for Configuration Management	United States Navy
10	Mil- STD-483	Configuration Management Practices for Systems, Equipment, Munitions and Computer Programs	United States Department of Defense.
11	MIL-STD-480A	Configuration Control-Engineering Changes, Deviations, and Waivers	United States Department of Defense
12	MIL-STD-1456	Contractor Configuration Management Plans	United States Department of Defense
13	MIL-STD-482	Configuration Status Accounting Data Elements and Related Features	United States Department of Defense
14	MIL-STD-481A	Configuration Control—Engineering Changes, Deviations and Waivers (short form).	United States Department of Defense
15	—	Fundamentals of Configuration Management by T. T. Samaras and F. L. Czerwinski	John Wiley & Sons, New York
16	—	Software Configuration Management by E. H. Bersoff, V. D. Henderson and S. G. Siegel	Prentice-Hall Inc., New Jersey

Configuration: the total engineering detail needed to construct, test, accept, operate, maintain, and logistically support a system.

Configuration Component: an element subject to configuration management.

Configuration Number: an alphanumeric designator, which distinguishes configuration components.

Baseline: an approved point of reference for control of future changes to design, construction, and performance of equipment.

Configuration Control: coordination, evaluation, and approval of all changes in the engineering product configuration described by the baseline.

Configuration Accounting: the reporting and recording of changes carried out to the configuration in question.

Configuration Status: the actual item configuration at a specified time, with respect to baseline or approved configuration.

9.3.2. Reasons for Configuration Management and Changes

Configuration management is used for the following reasons:[7]

1. To realize overall economy in terms of time.
2. To enhance the efficiency of the project.
3. To realize total economy in terms of money.

In general, three broad classifications of changes, either to design or production, are as follows:

1. *Changes type I:* those changes that are needed to rectify deficiencies.
2. *Changes type II:* those changes that improve logistic support or operational use significantly.
3. *Change type III:* those changes that minimize life cycle cost significantly.

There are various reasons for making changes during the design and production phases of equipment. During the design phase, the basic reasons for changes are as follows:

1. Availability of a part with better performance (due to modern technology).
2. To correct a deficiency (discovered either by analysis or test) in the proposed design.
3. To fulfill an objective more effectively.

During the production phase, the basic reasons for the changes are as follows:

1. The production people find it difficult to produce the product in the quantity required according to the existing design.
2. The manufacturing group has come up with an idea for change which will enhance performance significantly.
3. The manufacturing group has come up with substitute components, processes, or materials which will reduce costs.

9.3.3. Configuration Management Applications

Configuration management can be applied successfully during the life cycle of a product. It is especially applicable in the following four phases of a product's life cycle:[7]

1. Research.
2. Engineering.
3. Manufacturing.
4. Field support.

During these phases, the following problems can arise:

1. Ineffective control over changes made to equipment in the field environment.
2. Poor management direction on research projects.
3. Poor monitoring and control of changes during the research and engineering phases.
4. Poorly defined design objectives and concepts.
5. Poor standardization program and documentation during the engineering phase.
6. Inadequate communication with other groups during the engineering phase.
7. Poor managerial direction and review during the engineering phase.
8. Inefficiency in storage and retrieval of data during the manufacturing phase.
9. Poor control of revisions and incorrect data concerning the equipment produced during the manufacturing phase.
10. Wrong data explaining maintenance procedures.

These problems, however, can be corrected with the application of configuration management principles.

9.3.4. Phases and Baselines for Configuration Management

This section briefly describes the phases and baselines for configuration management. According to reference [7], configuration management divides the life cycle of a project primarily into four phases. These phases are (1) concept formulation; (2) definition; (3) acquisition; and (4) operation. With the exception of operational phase, these phases end at the baselines known as characteristics, functional, and operational. For example, the concept formulation phase ends at the characteristics baseline, the definition phase ends at the fundational baseline, and the acquisition phase ends at the operational baseline. In addition, the acquisition phase is further divided into the following two stages:

1. Design and development.
2. Production.

The design and development stage ends at the sub-baseline known as the *product* sub-baseline, and the production stage ends at the *operational* baseline. In the configuration management plan, baselines serve as checkpoints during the project life cycle. Thus, a new phase cannot begin until all questions and details raised in the earlier phases are satisfactorily resolved.

The configuration management techniques known as configuration identification, configuration control, and configuration accounting are applied during the life cycle phases of a project. The application of such techniques, however, varies to a certain degree, from one phase to another. Configuration identification is concerned with the identification of system items for the purpose of configuration accounting and configuration control. Identification is aided by technical manuals, technical specifications, engineering drawings, and related manufacturing data and information.[8] Similarly, configuration control is an orderly system evaluation through which baselines as well as systematic coordination are developed and approval/disapproval and implementation of (1) early baseline configuration, (2) systematic evolution of the system into the production baseline configuration, and (3) serviceability/maintainability changes in the final phase of the configuration management life cycle known as the operation phase, are made. Lastly, configuration accounting is concerned with the reporting and recording of changes made out to the approved configuration in question.

9.3.5. Configuration Control Board

The configuration control board, also known as the change control board, is composed of a number of technical and administrative project people, who are responsible for reviewing and assessing technical changes to the configuration item, after the approval of the baseline. More clearly, although it is not a voting board, its members present data and arguments concerning proposed changes. The ultimate responsibility for the decision-making regarding changes, is vested in its chairman. The structure of the board may vary from one project to another, or from one company to another, however, usually its members belong to the following groups:[9]

1. Manufacturing.
2. Engineering.
3. Contracts.
4. Quality assurance.
5. Production control.
6. Reliability and maintainability.
7. Purchasing.
8. Document/data control.
9. Accounting.

The principal functions of the board members are as follows:

1. To determine whether the proposed change is necessary.
2. To determine whether the cost requirements and schedule can be met without any difficulty.
3. To determine the feasibility of the proposed method of implementation.

Some of the questions usually asked by the configuration control board are as follows:[5]

1. Will the proposed change affect the service life of the configuration item?
2. What will be the resulting effect of the proposed change on the overall cost of the project?
3. Will the proposed change affect the safety of the operator?
4. Will the proposed change require special tools?
5. Will there be any problem with procuring parts?
6. Will there be any need for additional tests?
7. Will the proposed change make maintenance more difficult to perform?
8. Will the proposed change be effectively implemented by the manufacturing?
9. Will the proposed change affect the overall performance of the configuration item?
10. What effect will the proposed change have on spare parts?

9.3.6 Categories of Engineering Changes

The engineering changes are divided into the following two classes:

1. Class I engineering change.
2. Class II engineering change.

When the items, such as given below, are affected by the engineering change, the engineering changes, known as the Class I engineering changes, have an effect on the following:[10]

1. Safety.
2. Reliability specified in the contract.
3. Interchangeability.
4. Contract maintainability.
5. Contract delivery.
6. Contract specification.
7. Performance.
8. Contract weight and price fee.
9. Contract guarantees and test schedules.
10. Electromagnetic radiation hazards.

The configuration control board is primarily concerned with Class I engineering changes.

Class II engineering changes are those changes not included in the Class I category. Usually, the changes belonging to the Class II category are processed by the change planning group.

9.3.7. Benefits of Configuration Management

Some of the advantages of configuration management are as follows:

1. Reduction in overall cost.
2. Resources are channeled effectively.
3. Redundant effort is eliminated.
4. The resulting product is precisely identified.
5. Objectives are formally established.
6. The configuration management can be applied to large and small size firms, projects, or activities.
7. Accurate data retrieval facilitation.

9.4. VALUE ENGINEERING

Value engineering is concerned with the orderly application of established methods to identify the functions of a product or a service, and also with providing those identified functions at the minimum cost. The concept of value engineering was developed for application in the hardware area. However, in recent years, it has been applied in other areas as well. Historical evidence indicates that the application of value engineering concept has returned between $15–$30 for each dollar spent.[11] Therefore, the financial gain from the application of value engineering is very promising.

The development of the value engineering is credited to Lawrence D. Miles of the General Electric Company. In 1947 Mr. Miles was assigned a project to develop techniques that would reduce cost substantially, through material substitution or changes in production methods or design. Since then, several books on the subject have appeared, the American Society of Value

Table 9.2. Selective value engineering documents.

Item No.	Title	Produced By	Year
1	Value Engineering (Department of Defense Handbook 5010.8-H)	U.S. Government Printing Office Washington, D.C.	1968
2	Value Engineering (Engineering Design Handbook-AMCP 706-104)	U.S. Army Material Command Washington, D.C.	1971
3	Value Management: Value Engineering and Cost Reduction, by E. D. Heller	Addison-Wesley Publishing Company, Reading, MA	1971
4	Techniques of Value Analysis and Engineering, by L. D. Miles	McGraw-Hill Book Company, New York	1961
5	Value Analysis to Improve Productivity, by C. Fallon	Wiley, Inc., New York	1961
6	Value Engineering A Systematic Approach, by A. E. Mudge	McGraw-Hill Book Company, New York	1971
7	Value Analysis for Better Management, by W. J. Ridge	American Management Association, New York	1969
8	Guide for Application Value Engineering to Construction Industry, by A. J. Dell'isola	Mckee-Berger Mansueto, Inc., Washington, D.C.	1972

Engineers was formed in 1959, and a journal entitled value engineering first appeared in 1962. A list of selective documents on value engineering are presented in Table 9.2.

9.4.1. Definitions

This section presents selective definitions of terms associated with the value engineering discipline.[12]

Value Analysis: a systematically organized, creative procedure, whose objective is the identification of undesirables.[12]

Economic Value: the minimum cost to reliably provide the needed service or function at the desirable time and place with the required quality. (This is composed of use value, exchange value, esteem value, and cost value.)[13]

Value Engineering Proposal: a document that explains a recommended change that will result in an overall cost reduction (i.e. a change in design, approach, or equipment).

Value Objective: the same or better performance at reduced costs.

Value Engineering: concerned with the orderly application of established methods to identify the functions of a product or a service and also with providing those identified functions at the minimum cost.

9.4.2. Useful Guidelines to Start a Value Analysis Study

Before starting a value analysis study, the following information must be obtained:[14]

1. Starting and completion dates of the study.
2. Potential users of the end results.
3. Funds required for the study.
4. Requestors of the study.
5. Study definition.
6. Scope of the study.
7. Study approval body.
8. Study team.
9. Expected format of the study.
10. Involvement of organization levels.

9.4.3. Objectives of the Value Engineer

Objectives of a value engineer are as follows:

1. Identify and choose the most useful design procedures and techniques.

2. Choose the most profitable materials for the job in question.
3. Surface all product value reducing cost and quality factors.
4. Replace inadequate techniques, poor specifications, and inadequate design with better ones.
5. Choose the best and most economical product fabrication equipment.
6. Choose the most useful manufacturing processes.

The value engineer accomplishes the above objectives by seeking answers to the following questions:[15]

1. What is the component or the function?
2. What is the component's function?
3. How much does the component cost?
4. What alternative component will do the job?
5. What is the cost of the alternative?
6. Will the alternative be useful to improve quality?

9.4.4. Value Engineering Job Plan

The plan presents tasks or functions important to perform the study. Basically, the job plan consists of five phases as shown in Figure 9.2. These are[2]

1. The information phase.
2. The speculation phase.
3. The analysis phase.
4. The development phase
5. The presentation and follow-up phases.

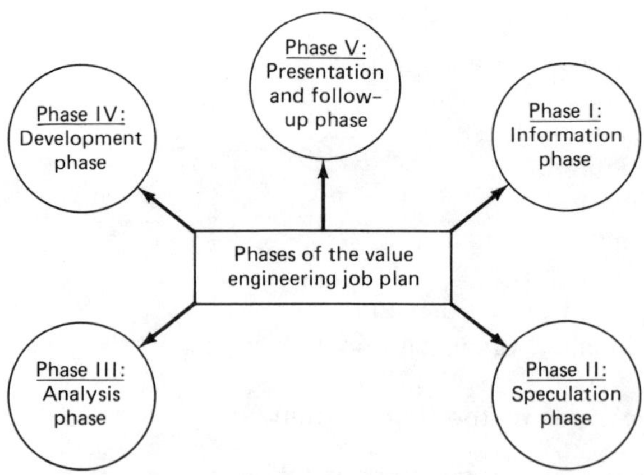

Figure 9.2. Phases of the value engineering job plan.

The first phase, or *information phase,* is basically concerned with seeking answers to task or function related questions such as (1) what is the component? (2) what does it do? and (3) what does it cost? This phase is concerned with working on specifics, defining the functions, gathering all the facts, and getting information from best sources. Examples of the type of information collected on a hardware project are

1. Information on cost.
2. Planning and schedules.
3. Quality control and reliability.
4. Information on procurement, manufacturing, and engineering.
5. Tooling and testing.

The *speculation phase* is concerned with seeking an answer to the question, "What alternative will also do the job"? In the *analysis phase* the answer to the question, "How much does it cost"?, is sought. The next phase, known as the development phase, is concerned with seeking answers to questions as follows:

1. What will fulfill the user's requirements?
2. What is required for implementation?

The last phase of the value engineering job plan is known as the *presentation and follow-up phase.* In this phase, answers to the following questions are sought (1) what was accomplished? (2) what is recommended? (3) what is the amount of money saved? and (4) who is the approval body?

9.4.5. Qualifications of a Value Engineer

This section briefly lists the qualifications of a value engineer. Some of these qualifications are as follows:

1. Ability to organize.
2. Sound technical background.
3. Aggressiveness and perseverance.
4. Creativity.
5. Ability to communicate.
6. Cost consciousness.
7. Basic training in value engineering.
8. Good education.
9. Drive to learn more about technology and industry.
10. Substantial experience in the industry.
11. Logical thought.
12. Ability to make decisions.

9.4.6. Advantages of Value Engineering

Some of the advantages of value engineering are[11]

1. It helps to identify areas that require attention and improvement.
2. It provides a technique to prioritize.
3. It provides a way to determine alternative solutions to a concern.
4. It provides a procedure to assign dollars on high-value items.
5. It provides a procedure to determine and quantify intangibles.
6. It provides a vehicle for dialogue.
7. It can be used as a process of prioritization to optimize resource allocations.
8. It provides a means for generating new ideas to solve problems.
9. It helps to increase the value of goods and services.
10. It provides a means to document rationale behind decisions.

9.5. NETWORK ANALYSIS TECHNIQUE

This section briefly describes scheduling techniques developed in the late 1950s, known as the *critical path method* (CPM), and the *program evaluation* and *review technique* (PERT). The program evaluation and review technique was developed for United States Navy by Booz, Allen, and Hamilton, a consulting firm, between the year 1956–1957. On the other hand, the critical path method was jointly developed by the Du Pont and Remington Rand Companies in 1957. Although both techniques have much in common, the basic difference is that the program evaluation and review technique allows, in its time estimates, explicit treatment of probability, whereas probability is not considered in the critical path method. Frequently, the critical path method is used in construction projects, and the program evaluation and review technique is emphasized in research and development projects. This section, however, concentrates only on the PERT.

9.5.1. Components of Program Evaluation and Review Technique

This section presents terms, symbols, and definitions associated with both PERT and CPM. Their symbols and descriptions are given in Table 9.3. Some of the terms associated with PERT are as follows:

1. Critical Path: the path through the network that composes several activities whose sum of duration times is the largest of any path through the network. If any delay in the completion of activities on the path occurs, the completion date of the entire project is delayed.
2. Critical Path Time: the sum of all activity duration times on the critical path.

Table 9.3. PERT and CPM symbols and their descriptions.

Item No.	Term	Symbol	Description
1	Event	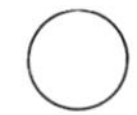	An inexplicitly identifiable point in time. In other words, a node in the network, representing the starting and/or ending of activities.
2	Activity		A required work component. The activity consumes time, money, labor, and materials.
3	Network		A combination of events and interconnecting activities that defines the logic of the project.
4	Dummy Activity		An imaginary activity with no actual time, which is used to preserve network logic.
5	—	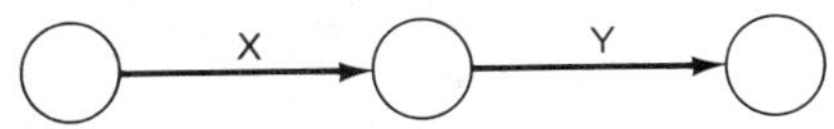	Activity *Y* cannot start until activity *X* has been accomplished.
6	—	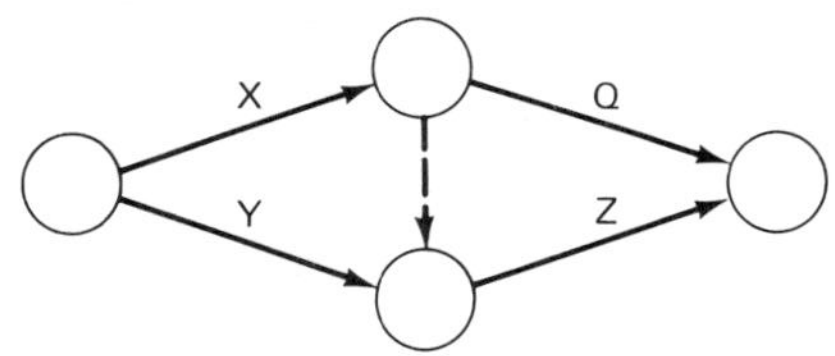	Activity *Z* cannot start until activities *X* and *Y* have been accomplished.
7	—	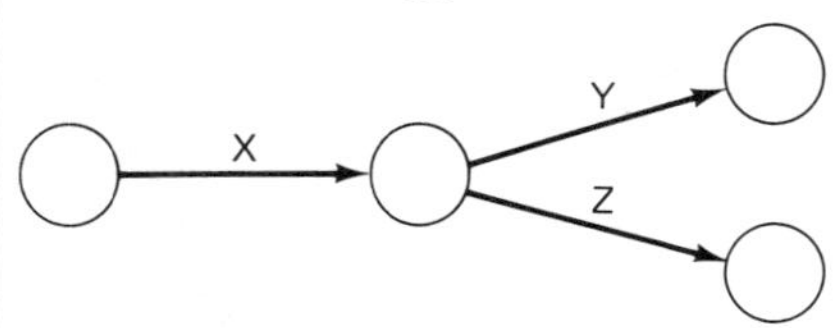	Activities *Y* and *Z* cannot start until activity *X* has been accomplished.

3. Expected Time: the mean time to complete an activity.
4. Optimistic Time: the completion time of an activity if everything works out ideally.
5. Pessimistic Time: the opposite of the optimistic completion time of an activity.
6. Most Likely Time: the most likely completion time of an activity.
7. Earliest Event Time: the earliest time an event can be reached. In addition, it represents the total of expected times up to that event.
8. Latest Event Time: the latest time an event can be reached without delaying the completion of the project.

9.5.2. Program Evaluation and Review Technique Logic

This method functions through the steps shown in Figure 9.3. The application of PERT has been found effective in the areas:[1]

1. Planning and smoothing resource utilization.
2. Obtaining cost estimates for proposed projects.
3. Scheduling projects.

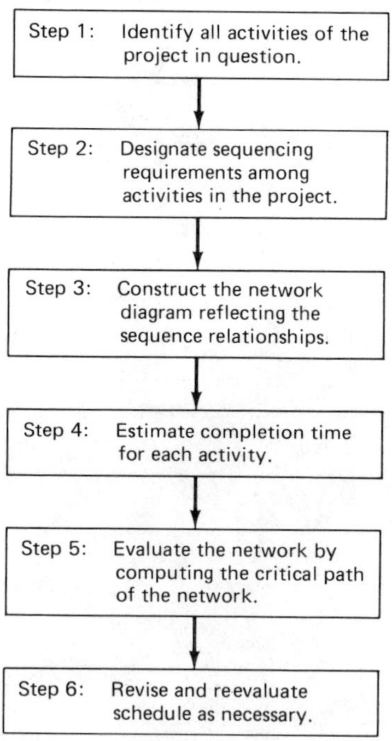

Figure 9.3. Logic of program evaluation and review technique.

4. Evaluating existing schedules and cost versus budget, as the work on project progresses.

9.5.3. Formulas Associated With PERT

This section presents two formulas to estimate the expected time for each activity and the variances of the activity times. It is assumed that time estimates follow a beta frequency distribution.

FORMULA I

This formula is used to estimate the expected completion time of an activity. Thus, the expected completion time of an activity is given by

$$T_m = (T_O + 4t_{me} + T_p)/6 \tag{9.2}$$

where

T_m is the expected completion time of an activity.
T_O is the optimistic completion time of an activity, if everything works out ideally.
t_{me} is the most likely completion time of an activity.
T_p is the pessimistic completion time of an activity.

EXAMPLE 9.1

Optimistic, most likely, and pessimistic times are 20, 24, and 31 days, respectively. Calculate the expected activity completion time.

By substituting the above specified data into equation (9.1), we get

$$T_m = [20 + 4(24) + 31]/6 = \frac{147}{6} = 24.5 \text{ days}$$

FORMULA II

This formula is used to calculate the variances of the activity times. The formula for the variance is

$$\sigma^2 = [T_p - T_O)/6]^2 \tag{9.3}$$

EXAMPLE 9.2

Optimistic, most likely, and pessimistic times are 12, 16, and 23 days, respectively. Calculate the value of the variance.

Substituting the specified data into equation (9.3), we get

$$\sigma^2 = (23\text{-}12)^2/6^2 = 3.3611$$

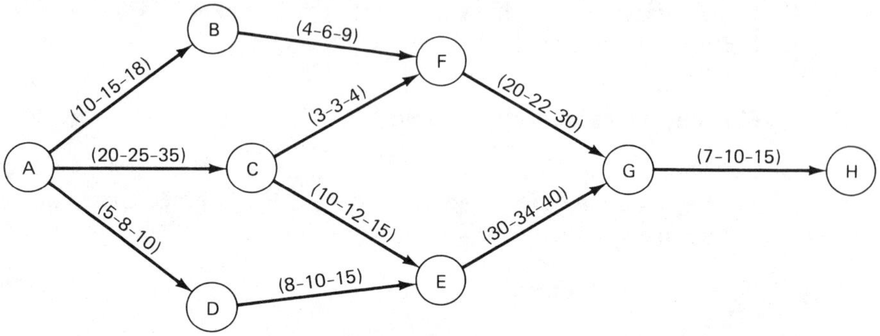

Figure 9.4. A PERT network.

EXAMPLE 9.3

A PERT network is shown in Figure 9.4. The numbers shown for each activity represent optimistic, most likely, and pessimistic times for each activity, respectively (e.g. for activity *AB*, the numbers 10, 15, 18 denote optimistic, most likely, and pessimistic times, respectively). Determine the critical path of the network.

Using equation (9.2), the expected activity completion time for activity AB is

$$T_m = [10 + 4(15) + 18]/6 = 14.67$$

For all the activities shown in Figure 9.4, the expected completion times are listed in Table 9.4.

The Figure 9.4 network is shown in Figure 9.5, along with the expected activity completion times from Table 9.4. This network has four different

Table 9.4. Expected completion times of Figure 9.4 activities.

Activity	Expected Completion Time
AB	14.67
AC	25.83
AD	7.83
BF	6.17
CF	3.17
CE	12.17
DE	10.50
FG	23
EG	34.33
GH	10.33

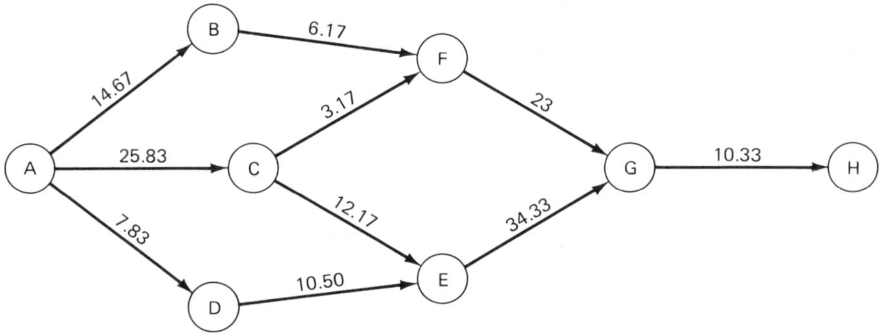

Figure 9.5. The PERT network with expected activity completion times.

paths originating at event *A* and terminating at event *H*. These paths are as follows:

1. *A-D-E-G-H.*
2. *A-C-E-G-H.*
3. *A-C-F-G-H.*
4. *A-B-F-G-H.*

For path *A-D-E-G-H* the total path time is = 7.83 + 10.50 + 34.33 + 10.33 = 62.99.
Similarly, for paths *A-C-E-G-H, A-C-F-G-H,* and *A-B-F-G-H* the total path times are 82.66, 62.33, and 54.17, respectively. Of the four values, the value for path *A-C-E-G-H* is the highest, therefore, this is the critical path.

9.6. SUMMARY

This chapter addresses four important areas: procedures for reliability assurance, configuration management, value engineering, and network analysis technique. In the procedures for reliability assurance section, the topics covered are designing for reliability, useful guidelines for designers and management, and reliability objectives in government contracts.

The next section describes the important aspects of configuration management, namely, the terms and definitions, the reasons for configuration management, the application areas of configuration management, the phases and baselines for configuration management, the configuration control board, and the advantages of configuration management.

Another important topic discussed in this chapter is value engineering. A brief history and selective definitions are presented, along with the guidelines to start a value analysis study, the value engineering job plan, the qualifications of a value engineer, and the benefits of value engineering.

Lastly, the program evaluation and review technique (PERT) is discussed. This method is described with numerical examples and their solutions.

9.7. EXERCISES

1. Describe a method that assures product reliability.
2. Discuss the histories of the critical path method (CPM), value engineering, and configuration management.
3. Describe the meanings of the following terms:
   i. Characteristics baseline.
   ii. Functional baseline.
   iii. Configuration control and control board.
   iv. Value analysis.
4. What are the needs for using configuration management?
5. Who are the members of the configuration management control board?
6. What are the qualifications and objectives of a value engineer?
7. What are the similarities and differences between the critical path method and the program evaluation and review technique?
8. A network is shown in Figure 9.6. Expected completion time for each activity is shown in the diagram. Determine the critical path of the network.

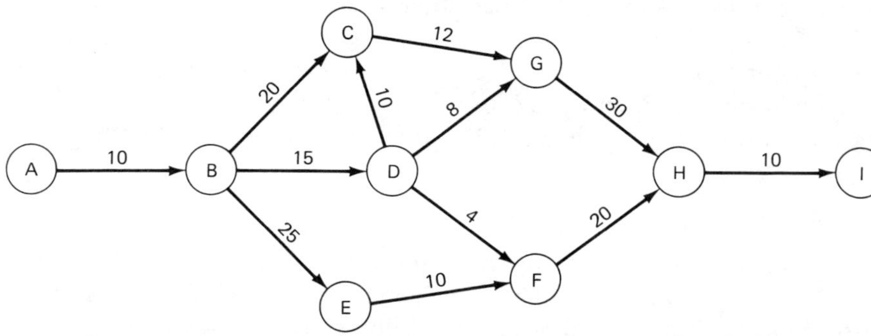

Figure 9.6. Network of a hypothetical project.

9.8. REFERENCES

1. Chestnut, H. *Systems Engineering Methods.* Wiley, Inc., New York, 1967. pp. 284–289.
2. Dieter, G. E. *Engineering Design: A Materials and Processing Approach.* McGraw-Hill Book Company, New York, 1983, pp. 459–462, 188–190.
3. Landers, R. R. *Reliability and Product Assurance: A Manual for Engineering and Management.* Prentice-Hall, Inc., Englewood Cliffs, NJ, 1963, pp. 275–276.

4. Chorafas, D. N. *An Introduction to Product Planning.* Cassell & Company, Ltd., London, 1967, pp. 259.
5. Samaras, T. T., Czerwinski, F. L. *Fundamentals of Configuration Management.* Wiley, Inc., New York, 1971.
6. United States Air Force. AFSCM 375-1, *Configuration Management During the Development and Acquisition Phases.* June 1962.
7. Hajek, V. G. *Configuration Management.* Industrial and Commercial Techniques, Ltds., 30 Fleet Street, London, 1968.
8. Memanus, E. J., Scianna, L. J. "Configuration Sets New Technique for Use in Configuration Management." *IEEE Transactions on Engineering Management,* Vol. 15, 1968, pp. 193–197.
9. Laine, M. J., Spevak, E. C. "Configuration Management." *Space/Aeronautics.* November 1966, pp. 74–81.
10. Feller, M. "Configuration Management." *IEEE Transactions on Engineering Management,* Vol. 16, 1969. pp. 64–66.
11. Demarle, D. J., Shillito, M. L. "Value Engineering," in *Handbook of Industrial Engineering,* edited by G. Salvendy, Wiley, Inc., 1982, pp. 7.3.1.-7.3.20.
12. Miles, L. D. *Techniques of Value Analysis and Engineering.* McGraw-Hill Book Company, New York, 1961.
13. Mudge, A. E. *Value Engineering: A Systematic Approach.* McGraw-Hill Book Company, New York, 1971.
14. Reigle, J. "Value Engineering; A Management Overview." *Value World,* Vol. 3, 1979, pp. 5–8.
15. Hayes, G. E., Romig, H. G. *Modern Quality Control.* Collier Macmillan Publishers, London, 1977, pp. 746–748.

CHAPTER 10

Reliability and Maintenance Manpower Requirements

10.1. INTRODUCTION

The maintenance of any system or equipment requires manpower. Since manpower is a costly life cycle parameter, the determination of the correct type and number of personnel for maintenance is a vital task which management must undertake to achieve effective maintainability within a given budget. Different maintenance philosophies, which include corrective and preventive maintenance, plus any related workloads that have a direct bearing on the actual maintenance time spent, must be evaluated. The maintenance factors involved are directly related to the reliability of a system or a piece of equipment. Therefore, it is essential that management has access to reliability and maintainability data for making the necessary calculations for manpower requirements.

10.2. MAINTENANCE PHILOSOPHIES INFLUENCING MANPOWER REQUIREMENTS

The military and most government agencies carry out corrective and preventive maintenance at three levels of organization. Although industry and commercial organizations have different maintenance philosophies, they all depend on manpower to carry out the required maintenance. The levels at which maintenance is done demand different degrees of skill at different time periods. The replacement of a pilot light or a plug-in module (once identified as having failed) can, in most cases, be carried out at the site of the system or the equipment, namely in the field. Repairs and preventive maintenance actions that require special skills and tools can normally be done only at higher maintenance levels, regardless of whether these maintenance levels are government, industrial, or commercial related.

10.2.1. BIT Contribution

The increased use of solid-state devices in practically all equipment effects reliability and maintainability, which in turn, effects the required manpower for maintenance. Another factor that must be considered is the contribution that BIT has made to the decrease in maintenance man-hours. Failure identification down to the individual parts level has not yet been fully achieved, due to design cost and consequently the requirement for more skilled maintenance personnel. Management dreams that one day no more maintenance personnel will be required during the life of an item, by using soft-fail features and self-correcting maintenance features. Such a dream, however, cannot be realized without prohibitive capital cost, and is not practical for the extended periods of time that equipment or systems are being kept in service. The workload (in terms of man-hours) is still extensive and the cost of skilled labor is increasing.

10.2.2. Data Needs

There are still a fairly good number of systems and equipment that plan to be in service for some years, which are not of the latest technologically advanced configurations. The failure rate trend and the maintainability features of such systems and equipment have been established over the user-years, and the required maintenance manpower may be calculated in terms of mean corrective and preventive maintenance times. This means that maintenance time may be used to establish the manpower requirements, and to adjust these requirements if, and when needed, during the life cycle. Failure rates for the lowest replaceable item in a system or a piece of equipment can be calculated in advance, using such guides as MIL-HDBK-217[5], the GIDEP[6], or other data sources. The various calculations may be cumbersome when large numbers of parts are involved, but automated data banks can assist. For each line or level of maintenance, the mean times can be evaluated in terms of the man-hours required to restore or keep an item operational, for a specified period of time. As will be shown, such calculations do not have to be complex, as long as they provide the necessary information.

Manpower requirements for older and existing equipment is based on the reliability data generated by the particular maintenance philosophy. Fairly accurate assessments of both corrective and preventive maintenance manpower requirements can be made. Time-consuming maintenance factors, which must be taken into account for such calculations, will be explained later in the text.

10.2.3. Performance Type Maintenance

The increased use of very complex solid-state devices has brought about some new ideas in maintenance and maintenance manpower. Routine preventive maintenance of older tube systems was carried out at specified intervals and corrective maintenance was down to the parts level. If equipment showed an increasing failure rate due to the presence of components with wear-out characteristics, then the components could be replaced before failure; this is true preventive maintenance. Replacement before failure requires that the failure distribution be known. From such a distribution, the time of replacement can be calculated. With modern type systems, however, there may be no wear out, and if there is, it occurs only after the part has been used for an extended period of time. Thus, this element of preventive maintenance is largely disappearing. Maintainers now claim that accurate maintenance time intervals cannot be determined for solid-state devices. In addition, complexity of both the hardware and the software make first-line or field maintenance impractical. Modules, such as printed circuit boards with extensive BIT and software structure, may show no clear failure pattern on which preventive maintenance could be based. Instead, what has been suggested, is to base maintenance on performance, rather than on time-related factors such as failure rates. Corrective maintenance is seen as a restoration of performance, and preventive maintenance as a reduction of performance failures through corrective maintenance. Preventative maintenance verifies system performance by end-to-end tests, which detect performance changes and possible performance failures, and also prevent error accumulation. Performance is a measure of the equipment stability, and no further maintenance action should be taken until another performance or functional test is carried out. Standards are to be used as a yardstick against which the results of the tests are measured.

The level of replacement, throwaway, or repair is governed by a repair and overhaul policy. The repair cost is to be taken into account for any decision.

To achieve a performance-based maintenance philosophy, a much higher skill level is required by maintainers. The repeated verification of equipment performance over a given operating time adds a considerable maintenance factor. Performance-based maintenance does not produce the failure rates, MTBF (Mean Time Between Failure), and MTTR (Mean Time To Repair), nor does it produce quantitative values for the calculation of availability. The idea being proposed is that any newly introduced equipment or systems will be performance-verified on almost a continuous basis in the beginning, and compared to the standard performance. After a certain period of operation, the verification of performance should be reduced to reflect the information obtained from previous performance tests. The word failure now means a change of the performance parameter against the standard or outside the acceptable limits.

10.3. MANPOWER REQUIREMENTS

How many man-hours will be required over a given time period to maintain a system or equipment will depend on the reliability, maintainability of the system or the equipment, and the required availability. A number of possible models can be used to determine of maintenance manpower requirements. Before discussing some of these models, some of the input requirements will be stated.

Corrective maintenance may involve some of the following procedures:

- Preparation for maintenance.
- Localization, isolation, and detection of failure.
- Disassembly for access to fault.
- Removal of fault item.
- Repair of fault.
- Installation of repair parts.
- Reassembly.
- Alignment and adjustment.
- Verification checkout.
- Reinstallation.

Preventive maintenance may involve some of the above procedures as well, but since no failure is involved, the repair and fault isolation procedures are unnecessary. Replacement of parts or modules may be a time consuming procedure.

In addition, there are other time-related procedures that require man-hours, which may or may not be productive maintenance man-hours. Productive maintenance man-hours are those hours spent keeping or restoring an item to a specified level of performance. For example, items removed as having failed, often do not have a failure rate at all. The so-called *no-fault-found syndrome*, is well-known to maintainers, and the number of such no-fault-found items can be considerable. This action does not contribute to maintenance productivity, and hence, is a waste of maintenance man-hours.

A logistic time delay may occur for the following reasons:

- Unavailability of spares for maintenance.
- Test equipment not ready or not available when needed.
- Waiting for transportation.
- Maintenance facility not ready.
- Maintenance documents not available when needed.

A logistic time delay is not considered as active maintenance time requiring manpower. Since it is impractical to fire maintenance personnel every time a delay occurs, the delay time must be accounted for in the calculation of required manpower. Another such factor that must be accounted for is administrative delay time.

10.4. RELATED WORKLOAD

Related workload factors, which are unrelated to basic corrective and preventive maintenance times are indirectly related factors such as travel, reporting, training, safety, ordering, etc. These related workload factors can amount to a fairly substantial time period, but care must be taken to ensure that any factors included in the maintenance man-hours calculations have a direct, not an indirect, bearing on maintenance, and are in line with the normal duties of a maintainer (not a clerk, supplier, laborer, or other administrative type). Adding workload factors that are not the responsibility of a maintainer results in padding of the required man-hours, and is usually done to increase personnel quotas. The list of related workload factors includes the following:

- Travel time to and from a maintenance site.
- Maintenance training time.
- Time to fill out forms for reports.
- Time to prepare orders for spares.
- Time to carry out safety checks.
- Preparation of test equipment.
- Time to give advice on maintenance matters.

Travel time may be used to obtain spares or test equipment for maintenance, from a location other than the maintenance area. For remotely located, unmanned equipment, the maintainer must travel to the site for corrective or preventive maintenance. Since maintenance training takes time, it must be figured into the available hours a maintainer has at his disposal. There are a number of different opinions about what constitutes training time, and how much of this time should be part of the total number of maintenance man-hours. In the majority of cases, the general training should take place prior to engaging in actual maintenance procedures and thus, should not be figured into the available time. Special training for newly-fielded equipment should be part of the accountable time. (This applies to on-the-job training time as well.) One particular government agency considered management training for maintainers, to be part of the mean maintenance time, since on large maintenance projects, a maintainer may have to act as a manager for a group of maintainers who require knowledge of management procedures.

Similarly, supervision time was counted as a time factor. It is suggested that such timings should not be added to maintenance time, nor should they be included in any other time.

The preparation of forms and the reporting of maintenance activities, as well as the writing of log books for information, are essential maintenance functions. There is a tendency, however, to overlook these time factors, especially when data is automated. Maintainers tend to report the absolute minimum of data, which takes as little of their valuable time as possible. Management and engineering normally require a fair amount of data for analysis and decision-making purposes. Regardless of how one compromises, data recording requires a fair amount of time. The ordering of spares for maintenance takes time, because part numbers must be known, descriptions must be entered, and quantities must be stated. Since not all this information is available in every case, it takes time to develop.

Not all maintenance actions require special safety checks, but equipment that is to be maintained with special safety features will require time for this. For example, equipment with explosive bolts require special safety checks before and after maintenance.

The preparation of test equipment for maintenance can take considerable time. Some of the modern test equipment require special set-ups that involve the selection of interface devices, the adjustment of stimuli, and interconnections to data recorders. There may also be an availability problem when only one digital spectrum analyzer is available when two are required at the same time.

A maintainer may also be called upon to advise or report on some specific maintenance aspects. Analytical reliability and maintainability studies often require additional information which has not been recorded. In such cases, experienced personnel can often fill in the gaps. There are other peculiar, time-consuming workloads that a maintainer may have to carry out. For example, a maintainer located at a site where he is not fully occupied, may carry out other non-related workload functions during the maintenance idle time. Consideration may have to be given for such special cases, even though these timings do not contribute to the actual maintenance time.

In order to arrive at the manpower requirement for an organization, management must have the mean times of the above-listed factors. It is also evident that much of the data needed are reliability-oriented. Manpower requirements should be stated as accurately as possible using all the factors that come into play when making calculations. The user is usually not concerned with how many people maintain his equipment, as long as he has an assurance that his equipment is ready and available for use at any particular time. Which data are needed by the user and which by the maintenance management must be considered when planning for manpower.

10.5. MAN-HOURS VERSUS THROWAWAY

Much has been written about the concept of *throwaway*. Most concepts are based on cost considerations. A simple replacement policy for certain types of replaceable units may sound attractive from a maintenance man-hour-saving standpoint, for first-line maintenance, and for achieving a fast turn-around. (There are other considerations that must be given to this concept.)

The cost of many specialty type, solid state devices, which have been tailored to meet some specific circuit function, are still very high when manufactured in small quantities. A throwaway module may be very costly, even though the total module price may be small compared to other modules. In addition, the fast obsolescence of some of these devices, and (especially those that have been tailor-made) the cost of manufacturing additional small quantities, makes them not economical. If such a device is used on a printed circuit board that has been identified as a field module but the particular device has not failed, it would be unwise to throw away the module with the functioning device, regardless of cost, unless enough lifetime spare modules were initially purchased to overcome obsolescence. Therefore, the idea of throwaway may not always be the best maintenance action to minimize maintenance man-hours and repairs.

10.6. NONPRODUCTIVE MAINTENANCE MAN-HOURS

The no-fault-found syndrome remains in modern systems just as it did in most older units. Failed modules that are replaced with plug-in modules for (especially at the first maintenance line) are often found to have no faults when tested at a higher maintenance level. The time spent to replace modules, which are not faulty but are believed to have failed, must be viewed as nonproductive maintenance man-hours. The reason why this nonproductivity occurs can be traced to a number of problems. Every maintenance management organization attempts to minimize this time, but experience shows that it still accounts for more than 20% of the total maintenance man-hours. This waste of man-hours reflects on the maintainer, who plugs the modules in and out and also on the person who tests the modules and ships them back to stores with the needed documentation. Every attempt should be made to reduce such instances of waste by determining the causes that lead to it.

10.7. MANPOWER MODELS

The maintenance man-hours expended must be considered. The time required to maintain a unit, namely the elapsed time, can be reduced by using additional manpower. An increase in manpower is an expensive cost factor, especially when highly skilled levels of maintenance are required to perform a particular maintenance task. Therefore, the maintenance time factor must be coupled with the cost in order to achieve a balance of manpower, time, and cost.

The mean maintenance time can be stated as

$$\overline{M_{ct}} + \overline{M_{pt}} = \overline{M_{mt}} \tag{10.1}$$

where

$\overline{M}_{ct}$ is the mean corrective maintenance time
$\overline{M}_{pt}$ is the mean preventive maintenance time
$\overline{M}_{mt}$ is the total maintenance time.

The exclusion of logistic and administrative delay times, which are separate management problems over which the maintainer has practically no control, should be noted.

The delay that may occur in maintaining a unit can be expressed as

$$\overline{M_{LDt}} + \overline{M_{ADt}} = \overline{M_{DLt}} \tag{10.2}$$

where

$\overline{M}_{LDt}$ is the mean logistic delay time
$\overline{M}_{ADt}$ is the mean administrative delay time
$\overline{M}_{DLt}$ is the total mean maintenance delay time.

The mean maintenance time and the mean maintenance delay time represent the total maintenance management time, which is the actual maintenance worktime, as well as the management time over which the maintainer has no control. Great care must be taken not to confuse these two parameters when calculating manpower requirements. Nevertheless, the total mean time could be expressed (keeping the above consideration in mind) as

$$\overline{M}_T + \overline{M}_{Mt} = \overline{M}_{DLt} \tag{10.3}$$

where

$\overline{M}_T$ is the total mean maintenance management time
$\overline{M}_{MT}$ is the mean management time.

The model of the manpower requirements for corrective maintenance only, may take on another form which includes some of the following parameters:

1. The maintenance man-hours (MMH) for each maintenance action or MMH/M_A. (M_A is the number of maintenance actions per given time period).
2. The maintenance man-hours for each month, or MMH/M_m. (M_m is the number of months).
3. The maintenance man-hours for each mission or operating cycle of a

system or an equipment, or MMH/O_c. (O_c is the number of operating cycles). (Here, the word operating cycle may mean the mission time, the flying time of an aircraft, the sailing time of a ship, the mileage covered by a vehicle, the number of cycles operated by a pump system, or any other such combination.)

These parameters can be expressed in terms of mean time values, and all may contribute to the mean corrective maintenance man-hours in the following way

$$\overline{MMH_c} = \frac{\Sigma\, \lambda_n\, (MMH_n)}{\Sigma\, \lambda_n} \tag{10.4}$$

where

λ_n is the failure rate of the nth item
MMH_n is the maintenance man-hours spent to maintain the nth item.

The total corrective and preventive maintenance man-hours can be expressed in a similar manner.

$$\overline{MMH_c} + \overline{MMH_p} = \overline{MMH_T} \tag{10.5}$$

where

$\overline{MMH_c}$ is the mean maintenance manhour corrective
$\overline{MMH_p}$ is the mean maintenance manhour preventive
$\overline{MMH_T}$ is the total mean maintenance manhour.

Each maintenance line or maintenance level may require different maintenance man-hours depending on the amount of maintenance work which is to be carried out at each organizational level. Specifically, care must be taken when calculating the maintenance time at the first maintenance line or level, because in some cases, maintenance is being performed by the operator rather than the maintainer. Adding the operator's time to the maintainer's man-hours could inflate the total man-hours, and thus result in more manpower requirements than necessary.

The above models do not include certain factors which have already been mentioned. The complete model may take on the following form:

- Mean corrective maintenance time— $\overline{M_{ct}}$ in hours.
- Mean preventive maintenance time— M_{pt} in hours.
- Mean corrective, plus preventive maintenance times,— $\overline{M_{ct}} + \overline{M_{pt}}$ in hours.
- Related workload time— M_{Rt} in hours.
- Travel time— M_{tt} in hours.

- Mean maintenance training time— $\overline{M_{trt}}$ in hours.
- Mean maintenance logistic delay time— $\overline{M_{LDt}}$ in hours.
- Mean maintenance administrative delay time— $\overline{M_{ADt}}$ in hours.

The total of all these factors represents an aggregate of maintenance work hours, maintenance management hours, and calendar time. Waiting months for a spare part does not add to the man-hour loading, but rather to the calendar time for repair. Many delay times are influenced by those management factors, and calculations should separate such factors from legitimate loads on maintenance.

A number of studies have indicated that the related workload time can vary from 20% to over 30% of the product corrective and the preventive maintenance times. Thus, the related workload functions must be chosen with care as they account for a fairly large number of maintenance man-hours.

The travel time factor varies widely from application to application. For remotely located items, travel time can be extensive. There is some difference of opinion as to how much training time should be allocated for each maintenance person. One particular organization allocates 10% of the corrective, preventive, and travel times sum to training. This appears high when compared with other organizations which allot only 2% of the sum to training, and do not take the travel time into account. (Here, travel time refers to the travel time to the training site, which may be the equipment location site or a school.)

The number of available man-hours/maintenance person must be stated in order to calculate, from the mean maintenance workload time factor, the number of maintainers required for a particular project. A number of different figures have been quoted, which vary from 1450–1550 hours/man/year. This small difference may appear to be insignificant, but when adding up values for a large fleet of 30 aircraft, each one requiring 3000 hours of maintenance/year, the difference is

$$\frac{90{,}000}{1450} = 62$$

and

$$\frac{90{,}000}{1550} = 58$$

This means that in one case, only 58 maintenance persons can carry out the job, and in the other case, at least 62 are required. A difference of four men/year, at a cost of about $30,000, means $120,000 extra pay. (The figures 1450 and 1550 were taken from personnel studies, which take into account such factors as leave, coffee breaks, sick leave, official holidays, etc.)

10.7.1. Cost Factors

As mentioned throughout the book, cost is one of the major management concerns. The cost of maintenance manpower takes a large part out of the total life cycle cost; therefore, the maintenance manpower requirements must be weighed against the cost. Various figures have been mentioned in the literature, which represent how much maintenance manpower contributes to the total life cycle cost.

It is obvious that maintenance, as a total entity, is a major cost factor. As a total entity it includes spares, test and support maintenance equipment, transportation and the handling, special facilities, and technical data. Manpower's contribution to cost of this entity, may range from 20%–35%. Therefore management must be aware of this cost factor. In a new equipment development project, management must evaluate the design, during the development, in terms of the impact on maintenance manpower, and may wish to determine the sensitivity of each parameter on manpower cost by using sensitivity analysis. A fixed manpower budget may force adjustments in MTBF and MTTR values to meet a stated expenditure level.

The maintenance cost factors can be established by calculating the salary and overhead cost of each maintenance skill level, which is required to carry out the maintenance of a particular system. The cost can be expressed within a given time frame, such as per maintenance action, per operating time or mission, per year, or any other time. Usually, various maintenance manpower skill levels are required on a particular job, especially when dealing with complex hardware/software maintenance actions. In this case, one could express the cost as follows

$$C_t = \sum_{i_1=1}^{n_1} m_{i_1}, \cdots, m_{n_1} + \sum_{i_n=1}^{n_n} m_{i_n}, \cdots, m_{n_n} \tag{10.6}$$

where

C_t is the total cost
n_1 is the number of skill level 1
n_n is the number of skill level n
m is the manpower cost factor. (i.e., total cost of men at each skill level).

For estimating the yearly corrective maintenance cost one could state

$$C_C = \frac{\text{MTTR}}{\text{MTBF}} h_o \, c_h \tag{10.7}$$

where

C_C is the total corrective maintenance cost
h_o is the operating hours of the system
c_h is the manpower cost per hour.

The cost values can be related to the total mean maintenance time. Adjustments will have to be made for the present values as they change with time.

10.8. EXAMPLES

The following example is for a Loran receiver of recent design.
The MTBF = 210 hours of the receiver, therefore
$\lambda = 0.00476$ failures per hour.
To calculate the mean corrective maintenance man-hours, given λ and the mean corrective maintenance time, equation (10.4) will give a

$$\overline{MMH_c} = 1.5 \text{ hrs/unit}$$

Similarly, calculation of the mean preventive maintenance man-hours can be made. This value is

$$\overline{MMH_p} = 0.85 \text{ hrs/unit}$$

The sum of these two represents the total corrective and preventive man-hours per unit. This value is

$$\overline{MMH_T} = 2.35 \text{ hrs/unit}$$

The related workload factor for this receiver has been established by maintenance management as being 26% of the MMH_T factor. This means

$$\overline{MMH_T} \times 0.26 = 0.611$$

The total then becomes

$$MMH_T + 0.611 = 2.96 \text{ hrs/unit}$$

Next, the training time must be added, which has been given 5% of the 2.96 hrs/unit plus the related workload factor, i.e., 2.96 hrs/unit. Thus the total time is 3.108 hrs/unit.

Travel time was very small for this unit because the receivers were located near a repair and supply area. The mean maintenance logistic plus the administrative delay time was

$$\overline{M_{LDt}} + \overline{M_{ADt}} = 1.03 \text{ hrs/unit}$$

This high value was due to the fact that not enough initial spares were purchased and that difficulties arose when ordering special-type parts which had long lead times. The total mean hours for a unit was now 4.13 hrs/unit, i.e., $\overline{MMH}_{\text{total}} = (3.108 + 1.03)$ hrs/unit.

The number of maintenance actions/given time period must now be calculated. For this example, a year was chosen. The number of maintenance actions required is

$$M_A = \frac{\text{hrs/year}}{\text{MTBF}}$$

$$= \frac{8760}{210} = 41.71$$

A single preventive maintenance action was taken, which had to be added to the above, making the total $M_A = 42.71$/year. The total mean maintenance hours/unit multiplied by the number of maintenance actions, will give the total mean maintenance man-hours required for one receiver/year, or

$$\overline{MMH}_{\text{total}} \times M_A = 176.39 \text{ hrs/unit/year.}$$

This means that one single maintainer may be able to maintain either 8.22 or 8.79 receivers/year, depending on the hours available, either 1450 hrs or 1550 hrs. In this particular case, 20 Loran receivers had to be maintained in a given site area. This meant between 2.28 and 2.43 man-years had to be available for this maintenance task.

10.9. OTHER CONSIDERATIONS

The question of when maintenance action is to be taken is important, as it will affect the manpower assignment. One individual organization suggested that for any nonredundant unit that fails, maintenance action should be taken immediately after failure. For a redundant unit with a MTBF greater than 2000 hrs, however, maintenance action could be taken the next day.

The ratio of corrective to preventive maintenance plays a part when cal culating manpower requirements. Various ratios were encountered, ranging from 70% to 95%; that is, either 30% or 5% represented the preventive maintenance time compared to the corrective maintenance time (not including any overhaul maintenance time). Preventive maintenance time depends on the type of equipment used. For older equipment, the ratio may be higher than for modern equipment.

Care must be taken when adding up the various time factors. For example, if a spare part is obtained, the maintainer usually verifies that the part he has obtained is the correct one. The time taken to do this, even if it is short, may be added to the maintenance time, but may also be added to the logistic delay time instead. The separation of preventive and corrective maintenance times may sometimes be difficult, because both actions may be carried out at the same time, although the records may not indicate this. Although a ratio of 1:5 has been suggested as reasonable, this value may change, depending on the type of equipment which is to be maintained.

10.9.1. Preventive Maintenance Action

The question of much how much preventive maintenance should be carried out and, with this question, how many man-hours are required, depends on the equipment failure rate over the life cycle and should be answered. Initially, more hours are required until the equipment has been in service for some extended period of time. In many cases, one encounters too much preventive maintenance. One reason for excessive maintenance is the so-called "fear factor," which arises from major and catastrophic failure incidents. Another reason for excessive maintenance is a human one, that is, there is a tendency for managers to have older maintenance personnel work on equipment in order to sustain their competence. Another factor is the existence of maintenance standards which spell out the maintenance hours required for a job, thus, the maintainers tend to spend maintenance hours as specified rather than try to improve on them. If the standard preventive maintenance time is unnecessarily high, the actual maintenance records will indicate a high preventive maintenance time. Thus, man-hours may be inflated.

10.10. MAINTENANCE MAN-HOURS TRENDS

Some trends in maintenance man-hours requirements can be stated. The increase in complex solid-state devices brings about a reduction in maintenance, particularly preventive maintenance. BIT will further reduce the number of maintenance man-hours, and with increased BIT sophistication, the reduction will continue. Moreover, soft-fail features increase mean time between repairs. Older equipment requires increased maintenance until their phase-out time. An increase in equipment types in every field and lack of standardization increases the maintenance man-hours in an almost linear fashion. Some reduction can be achieved, however, through revised maintenance procedures, improved and more reliable software, better training, and more extensive use of automation.

10.11. SUMMARY

There is a direct relationship between reliability and manpower requirements for maintenance, which must be considered by management. Because manpower is a costly slice of the entire life cycle cost, management must evaluate this requirement as best as it is possible, to reduce its cost. Here, the maintenance philosophy plays a major role.

Various models can be used to estimate the number of man-hours necessary to keep equipment in operating and useful condition. A simple example is given, which includes certain considerations that influence man-hour requirements. Where maintenance loads are large and costly, trade-offs against improved reliability and acquisition cost versus maintenance cost may be useful solutions.

10.12. EXERCISES

1. What factors contribute to corrective and preventive maintenance?
2. What factors contribute to maintenance delay times?
3. Two different pieces of equipment exist, one with a MTBF of 200 hrs, and the other with a MTBF of 390 hrs. There are six of the first type and ten of the second type. The ratio of corrective to preventive maintenance is the same for both, namely 80/20. The first piece of equipment is located near the repair sites, and the second is at a remote location (1600 KM away). (Assume other factors, such as travel and delay times.) Calculate the mean number of maintenance man-hours for each piece of equipment, and the total number of man-hours required for both.
4. Indicate the differences in maintenance requirements of a recently designed, state-of-the-art piece of equipment, and an older piece of equipment.
5. State some of the reasons for no-fault-found maintenance.
6. What are the cost drivers for equipment with low reliability?

10.13. REFERENCES

1. Blanchard, B. S. *Systems Engineering and Analysis.* Prentice Hall Inc., 1981, pp. 421–447.
2. Blanchard, B. S. *Logistics Engineering and Management.* Prentice Hall Inc., 1981, pp. 312–348.
3. Dhillon, B. S. *Systems Reliability, Maintainability Management.* Petrocelli Books Inc., 1983, pp. 134.
4. Reiche, H., Roy, Major J. G. C. "Reliability versus Sophistication." In *Proceedings NATO Operational Research Conference.* Ankara, Turkey, Sept. 1975.
5. Naval Publications and Forms Center. US MIL-HDBK-217, *Reliability Prediction of Electronic Equipment.* Philadelphia, PA.
6. US Naval Fleet Analysis Center. GIDEP, Government Industry Data Exchange Program. Corona, CA.

CHAPTER 11

Reliability and Maintainability Management Data Needs

11.1. INTRODUCTION

After having identified the activities of various managers in Chapter 6, one must now examine what kind of data that management needs to carry out its activities, when dealing with reliability and maintainability tasks. The data collected during the entire life cycle of a project, are essential pieces of information necessary for decision-making. That is, no valid decisions can be made by management without suitable data.

Information systems and data banks are essential management tools. Actions for corrective and preventive measures rely on available information, which has been collected and can be retrieved. The accuracy of information is important and therefore the type of information system or data bank is important. Although the conditions and the kind of data needed by management are different for each life cycle phase, the availability of data and the format of the data must be planned to meet specific requirements.

11.2. INFORMATION SYSTEMS

A number of military documents exist, which deal with the subject of data. For example, MIL-D-8706 deals with "Data and Tests, Engineering: Contract Requirements for Aircraft Weapon Systems;" MIL-D-26329 deals with "Data, Qualitative and Quantitative Personnel Requirements Information"; AFSCM 310-1 is titled "Management of Contractor Data and Report." There are an endless number of documents in the form of standards, which address specific data such as testing data, environmental data, quality data, safety data, maintainability and reliability information, life cycle cost data, and many others. Many well-known military handbooks cover reliability and maintainability data. It can be a major task for management to decide which of these data documents are important. Although different managers need different data at different times, all of them like to have information within easy reach, and in a format that can be understood and applied. The development of an information system for each individual manager and for each individual task may be a possibility, but such a system may be very complex, and would require a very large memory with many software programs. Such a system may not necessarily be compatible with any other information system. A major advantage of such a system could be that the format of any information could be uniformly structured. The large number of existing systems are not uniformly structured, and thus, are often confusing to any manger. For example, in one case the reliability information may be in the form of MTBF (Mean Time Between Failure), and in another case may be expressed as MTTF (Mean Time To Failure). Moreover, the inherent availability may be calculated in one system, and in another system the operational availability may be given. Such data are usually put forth without any further explanation and without any confidence level. Normally, a manager is unable to sort out such differing data, and using the wrong data can easily lead to false decisions because the subtle differences between certain quantitative values may not be understood.

It would be ideal if one could develop a single information system which could be used by all levels of management and which makes use of a standard structured format. In the paragraph on information system content which follows, one can see most of the data needed by various managers at various times, is fairly simple in qualitative and quantitative terms. The range of the total data used by all managers is very broad and, in almost all cases, interfaces with other data such as logistics, finances, and personnel. The integration of all the information needed by all managers, and the identification of all the common reliability and maintainability management parameters, may be possible, with the ever-increasing power of computers. Many organizations have undertaken studies to combine existing information systems and reduce the number of non-compatible systems. But at this time, no single, all-encompassing reliability and maintainability management information system exists.

Automation of any information system is essential and the retrieval of data from any system must be simple enough in order not to discourage managers from using it. Many wasteful man-hours can be spent if procedures for data retrieval are complex and require extensive training. Once a data system exists, its use should be mandatory to encourage data standardization; this is an aspect which can assist management greatly but is often overlooked.

There are a large number of existing information systems which have been specifically developed for reliability and maintainability information. There are other systems, however, which cover parts of reliability and maintainability information but were designed for maintenance or logistic purposes only. Most of these maintenance or logistic systems have been found to have some shortcomings in one way or another. This is probably due to the fact that the initial design of these systems was often directed towards different goals than those to which they were applied and later used. Seldom, if ever, was there any attempt to tailor the data to the management levels that required the information. (This was especially true for senior management levels.) The fault lies on both sides in that the system designers did not consult the managers, and the managers did not express their needs. Therefore, many reliability and maintainability information systems and data banks are useful to the engineering staff but not management; and the maintenance and logistic systems are useful to the maintainers and suppliers but not the operators, the decision-makers, or the analysts.

To change an existing information system from its original design to a design that tries to accommodate new applications, different users, and different management levels, is in most cases impractical and, if attempted, often results in an inefficient system. In time, advances in system development will help to overcome some of the problems mentioned above, but at this time, management (especially the senior level) is not well served by existing systems. The bottom line is cost, and all management information systems must address this need.

11.3. PURPOSE OF INFORMATION SYSTEM

The previous discussions on information systems leads to the question of what is the purpose of such a system? The purpose of an information system is to collect data which provide ongoing information used to analyze, evaluate, and assess the performance and the effectiveness of the operation; the maintenance and logistic support of the operation; the reliability and maintainability of the operation; the operation's manpower; and the life cycle cost of the operation. Therefore, anyone involved during the life cycle of an item wants to know, and have, certain information during the entire life cycle of a system. This in turn requires that an information system must be capable of providing the needed data at specific times to the individual manager. An information system cannot always provide ongoing data, in which case historical data may be useful and applicable for guidance, design, development, predictions, and future requirement assessments. More specifically, information collected from past and present experiences is applied to any new information for which only limited data are available, but where historical data can be presented. Any information is dynamic and its purpose must reflect the ever-changing needs of management. The growth and expansion of an information system is important if the system is to remain useful and valid.

11.4. INFORMATION SYSTEM CONTENT

What are the first steps leading to an information system? As simple as this question may sound, it is a most difficult one to answer. Most existing information systems were started because there was a need to generate some data. Rarely were the questions raised (1) who needs the information, (2) what information is wanted, (3) where is the information required? Without the answer to these questions, the development of an information system does not lead to a useful system. Each manager may require different information and different combinations of data at different locations, for entirely different reasons. At this time, however, it is neither practical nor cost effective to develop a single system which could be used by all managers. Integration of many existing systems may not be possible due to various reasons, as outlined before.

There is little to be gained by generating information that will not be needed. A survey of all parties requiring information is essential before establishing the kind of data that everyday work demands. It should be obvious that statistical mathematical information is not what the normal technician wants to know. It is also clear that the number of spares on hand in the supply system, or the inadequacy of equipment testability is of no interest to the president of a manufacturing concern; nor must the military commander of a field unit be given information on the mean time to repair for an equipment. The project manager will have a different view of his data needs during the conceptual phase, than will the life cycle material manager during the in-service phase. Although the total amount of data from all sources may be vast in scope, this book discusses specifically, the reliability and maintainability management data; therefore some of the following information may be important.

11.4.1. Specific Reliability and Maintainability Management Data

The Project Manager may require some of the following:

- General reliability data.
- General maintainability data.
- Availability data.
- System effectiveness information.

The life cycle manager will have to know much more detailed information about the life cycle phases that he is responsible for. The life cycle manager is not only concerned with what fails, how it fails, and why it fails, but also with how this information can be used to get an item back in service as fast as possible and how to keep it operating. Getting an item back in service and keeping it operating requires detailed reliability and maintainability information, maintenance information generated at all maintenance levels and lines, and specific details of the equipment or the systems effected. This data may take on the form of

- Type of unit.
- Serial number of unit.
- Part number.
- Modification number.
- Type of installation.
- Description of failure.
  a. Short.
  b. Open.
  c. Intermittent.
  d. Low output.
  e. No data transfer.
- Time of failure.
  a. Elapsed time.
  b. Mileage.
  c. Number of cycles.
  d. Time during mission.
  e. Time during checkout.
- Action responsibility.
  a. Design.
  b. Operator.
  c. Maintainer.
  d. Engineer.
  e. Manufacturer.
  f. Repair and overhaul.
- Maintenance data.
  a. Fault location time.
  b. Line replaceable unit (LRU) replacement time.
  c. Repair time.
  d. Checkout time.

This information must allow calculation of quantitative reliability and maintainability data.

The maintenance manager may want to know the following:

- Level and location where maintenance was done.
- Kind of maintenance carried out.
- Frequency of corrective and preventive maintenance.
- Downtime.
- Maintenance time.
- Logistic support times.
- Delay times.
- Spares availability.
- Test equipment availability.
- Manpower availability.

For analytical purposes, the analyst may need data on

- MTBF.
- MTTR.
- Mean maintenance free flight hours.
- Mean maintenance man-hours.
- Failure distribution.
- Cost of spares.
- Cost of manpower.
- Cost of software.

A commander in the field may use

- Manpower availability.
- Equipment availability.
- Equipment location.
- Readiness status.

Life cycle cost data may not be easy to come by during the initial phases of the life cycle, and the available data may not be accurate because of many factors which come into play when introducing a new item. The total life cycle cost can be given only after the item has been disposed of, but it is preferable to make use of as much cost data as can be generated, rather than to make guesses from so-called experiences, which often have no relation to the question at hand.

11.4.2. Data Deficiency

Regardless of what kind of information is collected, typical deficiencies in reliability and maintainability management data will always exist. These may be stated as

- Lack of standardized terms and definitions.
- Inconsistent data reporting.
- Time-consuming data retrieval.
- Doubtful validity of data recorded.
- Inaccurate time records.
- Lack of cause for failure identification.
- Poor item description.

The first two are the major deficiencies which management is faced with. Little, or not time is normally available to review all data before using them. For analysis purposes, this lack of time presents a problem which can only partly be overcome by extensive reviews and additional discussions.

11.4.3. Data Deficiency Trends

Various data deficiencies have been discussed briefly in the above paragraph. One particular area of concern has been the data requirement for the latest integrated circuit designed systems. For many years, reliability and maintainability predictions ("if-then statements") have used the failure rates of all parts in any reliability calculation. Management specified that such failure rates be taken from the latest US MIL-HDBK-217[7] version and be applied to discrete parts, as well as complex integrated circuits, such as microprocessors. The handbook lists generic failure rates, together with modifiers and applicable models, thus allowing the calculation of reliability for parts equipment, and systems with reasonable accuracy. The modifiers, which must be used, are to reflect the eventual field conditions and the stresses, to which each part is subjected, under maintained field conditions. Unfortunately, this kind of data is often unavailable in the early life cycle and, therefore, many assumptions must be made and provided by management in order to obtain quantitative reliability and maintainability values. An understanding of these data is important to management when planning its initial, integrated logistic support requirements.

With the increasing complexity of systems, the number of individual data points, which are required for any reliability and maintainability calculation increases, thus making this kind of exercise very time-consuming and costly. Automated data bases do assist and speed up the calculations, but the various modifiers and stress levels, which must be applied, must be selected manually and fed into the model for each part, unless all parts are assumed to make use of the same values. Shortcuts are often used but reflected in the accuracies of the final results. All this effort is not encouraging to engineers and management, who are interested in timely, economically, and accurately eveloped data for decision-making purposes.

When circuits were simple and active elements could be used as a basis for evaluating reliability, not much engineering time and cost was involved. Special graphs, showing reliability in terms of complexity, active elements, and use factors, gave a ball-park failure rate, and this data was quickly available to management. These graphs were one method of reducing the endless part count and its associated calculations.

Now that discrete parts have been replaced by ICs, and the functions each device can perform has multiplied manifold, the number of discrete parts have been reduced. Further large-scale integration will continue over the years and further reduce the part count. But the trend to obtain more valid and meaningful reliability data is becoming more evident and difficult. Each new device requires a good understanding and a knowledge of the internal construction, and circuitry and device characterization is essential. Much of this detailed information is not made available by manufacturers, and the trend towards avoidance of proprietary information release is increasing. This applies especially to custom-tailored devices which, in many new systems, comprise a large portion of all devices used. Where then, should the needed data come from for any reliability calculation?

In a paper, "Are Prediction Methods Outdated?" (and in a few others) it is suggested that there may be other methods to improve and speed up predictions.[4] These papers emphasize the need to get away from the usual failure rate data part count, and the need to reduce the time and cost of such exercises in the interest of better management. A circuit must perform certain functions in an equipment or in a module such as on a printed circuit board. Usually, a large number of different functions can be identified, which make up the total functional capability of a piece of equipment. Certain functions may be simple, others more complex, and a single module may contain a single function or a number of different functions. The functions may be switching, digit processing, amplifying, or input/output controlling. If one was able to determine all the functions designed into the various modules, which are required to operate the equipment, and was able to look at the environment in which these functions have to operate, one would be able to relate these functions to failure rate data instead of parts failure rates. Functional block failure rate data would greatly assist in cutting down the cost and the time of analysis and, hopefully, improve reliability assessment.

The number of functions in one piece of equipment will certainly not be as large as a part count. For example, a digitizer module could be evaluated as a function instead of adding up all failure rates of the devices, resistors, capacitors, diodes, and other parts. If such a function fails, this data could be used to calculate the function failure rate under certain operating conditions. The listing of function failure rates forms the basis for the reliability data bank. With digital equipment, many functions are often repeated, thus reducing the calculations, as long as the functions are the same or similar. The Government Industry Data Exchange Program already contains many data which could be used to reduce the number of calculations. Engineers and managers would do well to examine this kind of approach of data collection and reliability evaluation.

11.4.4. Reliability Growth Data

Extensive testing is normally carried out in order to obtain the needed reliability data from which a reliability growth prediction can be made. Because many of the large systems under development are not produced in any large quantity, and because timing of the introduction into the field is often critical, tests to establish a growth trend (which are to meet a stated MTBF) cannot be carried out economically, in the limited time frame. Abbreviated or accelerated test data, if available at all, may often not reflect the reliability of the system under actual field condition. An interesting solution to certain types of reliability growth application problems was suggested in a paper entitled, "An Approach to Reliability Growth Without Dedicated Testing: A Case History and Some Results."[5] This paper describes how the needed reliability data can be collected during a planned engineering development project, how any problems can be fixed and how the data can be analyzed. The data can be obtained without dedicated hardware or dedicated reliability growth testing. This kind of approach can be advantageous if the system under development encounters a similar environmental profile during its development as it does under fielded condition.

The longer the development time for the total system, the more reliability data can be collected. Large systems, which require hardware integration, and interface the various developmental tests of all units making up the system, can take a few years to develop. If one operated all the units of the system continuously during the development cycle, some units would accumulate a fairly large amount of test time. The total test time for all units together will be large.

One of the key acivities, which must be established by management, is failure reporting. Here, the rule which has been found to be the best is to record every incident that occurs. Any anomalies are to be reported, which means any deviation from what is expected must be reported. All problems would then be captured, even if this means that some data may not be relevant. The record of these data should then include everything subjected to development testing, even if some units are not the direct responsibility of the developer (e.g. government-furnished equipment, or a unit on loan from another source).

Because there is a definite need to obtain as much data as possible, it is better to collect everything rather than only parts of the incidents. It often turns out that the unimportant data becomes a problem that requires correction. Once all data have been collected during the development period, they must be analyzed and placed in the categories of relevant, non-relevant, and not applicable information. The data can then be further divided into the kinds of relevant incidents, such as defects of parts, workmanship problems, design defects, operating and maintenance procedure problems, and others. Once this has been done, a reliability growth model can be chosen.

The selection of a suitable model is important. For example, the Duane Model, which is based on the cumulative average, is influenced by the "old" data points and this is reflected in the final accumulated data plot. But this model can calculate the last MTBF data (referred to as current MTBF), and this will indicate the slope at the end of the growth curve and the current MTBF value. *MILD-HDBK-189, Reliability Growth Management,* shows another model (AMSSA) which can handle continuous changes in slope of the reliability growth curve by restarting calculations when such changes occur.[8] This model is more complex and not as easy to understand, but it avoids the influence of the "old" data points.

This model also has a number of features that should be of interest and of value to engineers and managers alike. First, it is a method that in comparison to other analytical methods, gives superior quantitative reliability results. It leads to the reduction of, or even the elimination of, formal reliability growth tests. Moreover, it provides the possibility of collecting more and better data for analysis and for any needed corrective action. The amount of data collected with this model can be of greater magnitude than the amount collected from formal tests. It is a simple technique, which can be understood by all management levels and whose results can be displayed in any readable form. The results give management a better assurance of the fielded equipment reliability, and also eliminate failure mechanisms early on. It should be kept in mind that the data required to establish the initial MTBF value for any of the growth trend analyses is important. Both MILD-STD-1635 and MIL-HDBK-189 strongly recommend using reliability history data of similar systems, but discourage choosing an arbitrary starting MTBF value, such as 10%.

11.5. DATA RANGE AND IMPLICATION

From the above paragraphs it is evident that a wide range of data is required to meet all the needs. A careful review of all stated requirements may reveal that there is some duplication of the information wanted by different management staffs. The integration of such common data into a single system, accessible to anyone, and in the format which can be used and understood by all users, is the ideal that all designers of management information systems strive for. Such a system would be much more powerful in terms of available and consolidated data, but would also allow a closer link between reliability, maintainability, and such factors as cost and logistic information. Programming of such a system would be easier and would also achieve specific standard format output. The attempt to relate engineering reliability and maintainability data with logistics, when both are using different information system concepts, is very difficult. For example, an engineering maintenance information system may be equipment/system-oriented, identifying each record with a particular item, such as a radar from a frigate. The logistic system, which may be supply-oriented, may identify the subject only by the commodity name, such as radars. There may be no practical way to extract data from both systems to come up with an answer, which may involve both maintenance engineering and supply, unless some additional information is made available. For example, it is important to a maintainer where a particular part is used, even if such a part may be common to more than one piece of equipment. For equipment, in which such a part is used, the failure rate may be different due to different circuitry and application. The maintainer certainly has to assure himself that the part fits exactly into his equipment, and that he has available to him, the exact spare parts for the maintenance of his equipment. He is not interested at all in how this same part fits into an entirely different piece of equipment over which he has no control. The supply information system cannot separate this part from any other part, as its use is recorded only by the name of the commodity, such as radar, and not by the specific piece of equipment for which the maintainer needs the part. Because the spares calculations are based on failure rates, each radar using the same part may have different spares requirements. Thus, the idea of a single supply stock for a part is attractive from a logistic viewpoint, but leaves a wide information gap open for the maintenance engineering manager, who is only interested in a particular part. Moreover, a single supply stock also makes it impossible for the maintenance engineering manager to state, with any accuracy, the maintenance cost for his radar, because only the total inventory part cost for all radars is recorded.

The reason so many management information systems exist side by side in one organization stems from the fact that each range of data requirements was accommodated within many different systems, and no attempt was made to initially coordinate them. Each individual user began developing their own system, resulting in many incompatible systems. In addition, many systems were developed before new management philosophies pressed for different data than was originally wanted. Changes to the older systems are difficult to achieve with major compromises.

11.6. SOFTWARE DATA

The requirement for reliability and maintainability software data for management purposes is fairly new. Data are needed during the first three life cycle phases. The data may include, but not be limited to, the following items:

1. Name of software and latest revisions.
2. Short description of software function.
3. Software input and output data.
4. Software assumptions and restrictions.
5. Software design alternatives.
6. Short history of revisions and reasons for them.
7. Test methods.
8. Test data and test results.
9. On-going software support.
10. Software maintenance concept.
11. Failure conditions and corrective actions.
12. Failure and error probability.

11.7. INFORMATION USERS

The question of who needs data has been answered in part. During the entire life cycle, various decisions have to be made by management at all levels. The first person to deal with any reliability and maintainability problem is the project manager. During the conceptional phase, he will be required to deal with many subjects, review options, evaluate the impact that any changes may have, and make decisions. The manufacturing stage is followed by the in-service phase, during which the life cycle material manager has responsibility. Extensive use is made of any information eminating from the information systems. The life cycle material manager, and his staff, must assure the operator and the user that the equipment is ready and available when needed for the purpose stated, and under the conditions specified. The supply, personnel, and finance managers all require certain information for their tasks. This kind of information may not be in the form of reliability and maintainability data, but much of it may be related to these factors. Managers require an ever-increasing amount of data, and at the senior management level, data are required to plan policies, strategies, and finances for the department or organization.

If one allowed the distribution of all available data to all data users, a flood of computer run-outs would fill each desk. The majority would not know what to do with all these data, and would not even have the time to read a small part of them. Many would not be able to separate what is of value to them and what is not. Once it has been determined what information each manager needs, and for what reason, the distribution of only the data that applies directly to a manager or an organization should be released to them. There will be a considerable difference in the amount of data that an engineering manager can use, compared to what a president of engineering requires.

11.8. INFORMATION TIMING

The question of when and where the information is needed must be answered at an early stage. There is not much to be gained by sending two hundred computer pages, at the same time, to two engineers, who are sitting side by side in an office, and who both work on the same project. There is not much sense in mailing to each maintenance technician in the field, a stack of maintenance data. Generally, it is better to deliver more direct and useful information. Although there may be some who like to have their desks filled with computer papers, most like to see only an absolute minimum. Unfortunately, managers of information systems love to show everyone what great amount of data have been generated by them, and therefore make sure that everyone gets as many data as possible from their system. The need to time data distribution has seldom been addressed.

How often data is required will depend on the type of information required for specific tasks. Each manager may have different milestones for his task targets and for possible briefings. Most managers follow some sort of schedule. There is a tendency to request data on an almost daily basis for any newly introduced equipment. In such cases, the unfamiliarity generates a desire to continuously monitor events for a reasonable time period. This practice, however, is not based on any scientific information and leads to increased cost and waste of time, and is not much benefit to the user. The timing of the release of data should therefore be governed by the task milestones that must be met, and any schedules that an organization must follow. This means that data must become available when needed, and not when they are ready by the computer.

11.9. INFORMATION FLOW

If management wants to take action on any problem that arises, then the data leading to this action decision must follow a certain flow pattern. It is important that the handling of data, and the monitoring of any actions resulting from the data, be followed in a consistent way. Fig. 11.1 outlines the decisions and action flow requirements.

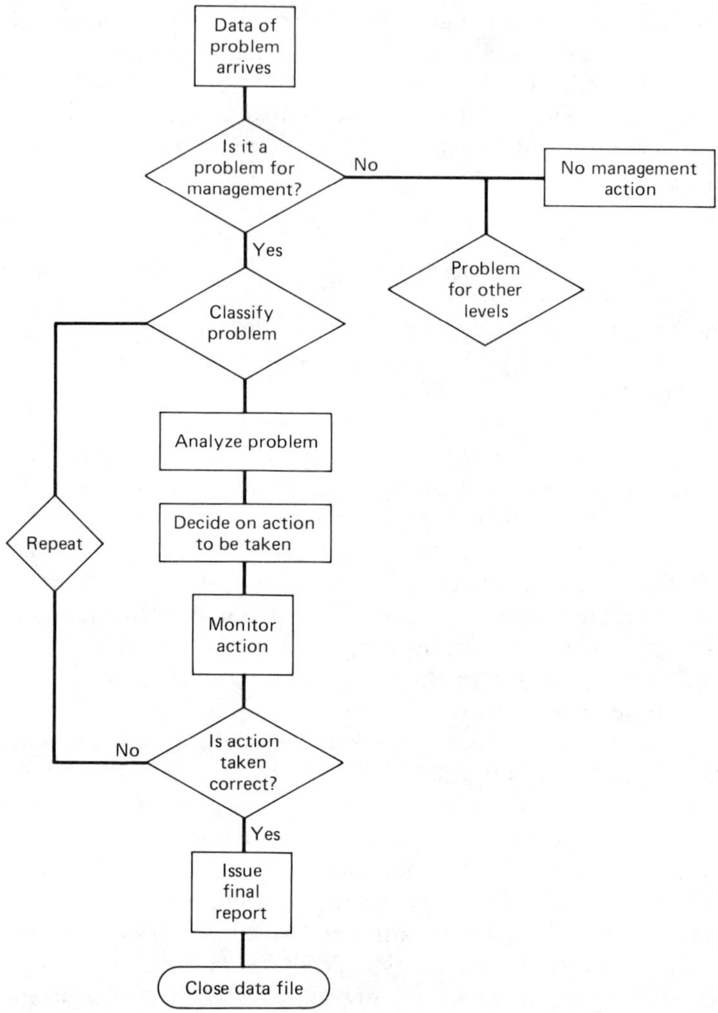

Figure 11.1. Decision and action data flow.

11.10. THE COST OF DATA

One factor that management must take into account when developing any data or information system is the cost. If no system exists from which management data can be retrieved, any development cost must be assessed against the use value. If reliability and maintainability data is not needed for all projects, it may be more cost effective to make use of external data sources, rather than developing a data system. There are many such external data sources available, which can be consulted with limited funds. An example is the data contained in MIL-HDBK-217. Software can be purchased or rented which allows calculations of failure rates and other factors contained in the handbook. Changes and amendments are made to these programs when required, and the user of a rented program does not have to spend additional funds on his own to make changes, unless he owns the program. An existing information system may have to be changed to meet new management concepts. In this case, the development of changes may be very costly. Changes to data are unavoidable with time if the data are to remain valid for management use. If user requirements change quickly, careful consideration must be given to the time frame to decide when changes should be made, keeping the cost factor in mind. A good example can be quoted. A manufacturer of certain electronic hardware for industrial applications wanted to keep track of maintenance, repair, and cost factors in order to determine the reason for certain reoccurring faults. An information system was developed for the manufacturer, for a cost of $120,000. The system was hardly installed when the company decided that a new maintenance concept and a new logistic support plan had to be added to the data. Thus, the entire system had to be redeveloped.

11.11. DATA RENTAL

Many organizations have their own reliability and maintainability data banks, but continuous changes must be made to any bank in order to keep the data up-to-date. The update is not a minor management problem and can be extensive if the data bank includes a large amount of high-technology information. A good example is the fast-changing field of failure rates for integrated circuits. The Reliability Analysis Centre at Griffiss Air Force Base, N.Y., publishes such information, and updates it continuously. Specifications and standards are also changed often by revisions, amendments, and completely new issues. Information of any changes or planned revisions are listed in some publications. A good update is given with every published review.[6] It is not a minor management task to keep informed about changes on all the reliability and maintainability data, which may be required for a particular project, but management must be aware of them, and have the new information available for decisions.

One possible solution to this problem is to buy or rent a data bank, which is kept up-to-date by the producer of the bank. Such a data bank is not only revised automatically, if and when needed by the producer who sells or rents the bank, but the complete data, which remains updated, can be displayed and read at any time. The availability of this kind of data is important and eliminates the problem and the need for a search of the latest existing data, amongst endless other data files.

A number of data banks can be rented or purchased. The information can be on a tape, floppy disc, or any other memory device. The data can be displayed on a terminal for reading and copying. Retrieval of the latest data is normally ensured by limited expenditures for their purchase and rent. Reliability journals list a few sources of supply for such data banks, which automatically provide data on the latest documents, failure rates, procedures, models, tests, and other aspects.

11.12. DATA OVERVIEW REQUIREMENTS

No manager can be familiar with all the reliability and maintainability data that he requires for any of his particular projects. He usually relies on others to keep him informed. There is a definite need of a data overview as a minimum at that level. Abstracts of data may be sufficient. If this minimum amount of data can be made available, it can assist a manager in his efforts. The presentation of such minimum data can be simple and without too much detail, but the updated overview will allow for more effective decisions. At this time, no such simple overview package is readily available to management. Requests for such data may encourage others to assemble this on any project for management.

11.13. SUMMARY

An information system plays a major role in the decision-making process. The purpose of any information system is to meet the needs of users, and therefore the content, format, and timing must be considered when developing any system. Not all users require all information and selected data distribution at appropriate timings is essential for an effective information system. The promotion of an integrated information system is encouraged, but benefits can only be realized if all levels of management indicate their requirements.

11.14. EXERCISES

1. Management requires data for decisions and actions. Envisage an engineering manager at the level of vice-president of engineering of a large manufacturing concern. What kind of reliability, maintainability, and other data would be required for his job?
2. Develop a simple data flowchart that outlines the information needed at each of the four life cycle phases.
3. This chapter mentions the importance of the accuracy of data for management. How can data integrity be built into an information system?

11.15 REFERENCES

1. Reiche, H. "Technical Data Package for Procurement." *International Journal Microelectronics and Reliability*, Vol. 23, No. 23 (Pergamon Press), pp. 509–518.
2. Blanchard, B. S. *Logistic Engineering and Management*. (Prentice Hall Inc., 1981), pp. 148–149, 291–296.
3. Dhillon, B. S. *Systems Reliability, Maintainability Management* (Petrocelli Books Inc., 1983), pp. 255–257.
4. Reiche, H. "Are Prediction Methods Outdated?" *Reliability Review* Vol. 3, March 1983, pp. 8.
5. Bentz, R. W., Hutchinson, Maj. L.T. "An Approach to Reliability Growth Without Dedicated Testing: A Case History and Some Results." In 1984 Proceedings Annual Reliability and Maintainability Symposium, pp. 458–463.
6. Green, K. "Government R and M Activities," in Reliability Review—A News and Opinion Quarterly, all issues (Reliability Division of ASQC).
7. Naval Publications and Forms Centre. US MIL-HDBK-189, *Reliability Growth Management*. Philadelphia, Pa.

CHAPTER 12

Reliability and Maintainability Design Review and Design Change Processes

12.1 INTRODUCTION

The total "life of a system (or equipment) may be viewed in terms of the time period extending from the initial planning (which takes place as the result of the identification of a requirement), to the eventual elimination of the system from the operational inventory. In the interests of efficient engineering and maintenance management, the system life cycle is frequently engineering and maintenance management, the system life cycle is frequently divided into four discrete phases and definitions: conception, acquisition, in-service, and disposal (see Chapter 6).

Between the two extremes of the system life cycle, material managers will be called upon to undertake or participate in design reviews, evaluate proposals to change, and deviate from, or waive specific aspects of the design or configuration of the system. Whether the purpose of a proposed change is to improve the system's availability, dependability, capability, and/or to reduce its acquisition or operating and maintenance costs, a careful analysis of the potential impact on reliability and maintainability is essential.

The purpose of this chapter is to furnish an understanding of the design review and design change processes, with emphasis on reliability and maintainability.

12.2. DESIGN REVIEW OBJECTIVES

Design reviews are conducted to facilitate decisions on the design of configuration items or aggregates of configuration items. Each review should result in a decision either to accept a given set of design parameters or to undertake further development and analysis activities. The major specific aims of the design review process are (1) to evaluate design progress, (2) to identify and resolve producer and customer risks associated with design adequacy, cost, and scheduling, and (3) to confirm that design details are compatible with contract functional specifications.

12.3. TYPES OF DESIGN REVIEWS

12.3.1. System Requirements Review

The customer normally defines and specifies a level of cost-effectiveness—based upon his operational requirements and budget—with which a system will be expected to meet. Effectiveness requirements will in turn be dependent upon availability, dependability, and capability. Reliability and maintainability are the essential ingredients of availability and dependability, and a specified change in capability influences system complexity, and thereby impacts upon reliability and maintainability.

Cost-effectiveness definitions and specifications may appear in a contract, a request for proposal, an invitation to tender, or any similar document designed to solicit and evoke a response from a contractor. The contractor then defines and documents his interpretation of the functional requirements for the system, which best meet the cost-effectiveness criteria established by the customer. To facilitate an evaluation of the contractor's proposal, one or more system requirements reviews must be conducted.

If a system is of mature design, and its cost-effectiveness has been proven and documented on the basis of another customer's experience, a single system requirements review may suffice to confirm the system's acceptability in the current circumstances. Alternatives to the "off-the-shelf" purchase of a mature system can range from specified minor or major design or configuration changes, to an entirely new system, designed and developed expressly to fulfil customer requirements. In the latter instance, several system requirements reviews are likely, before an accord is reached on the acceptability to the customer of the contractor's definition of system functional requirements.

12.3.2. System Design Review

The final definition of system functional requirements, as established by the system requirements review, must be translated into a proposed system design by the contractor's system engineering organization. The system design process involves the allocation of functional specifications to various segments of the system, normally identified at the configuration item level of indenture. An essential part of the foregoing process is the allocation of reliability and maintainability values to configuration items. The process may be based upon reliability and maintainability data representative of actual configuration items, upon theoretical projection and analysis, or upon an appropriate combination of the two.

A system design review's purpose is to determine that the proposed methods for attaining the required functional characteristics are logically sound, and promise to be as effective and economical as possible, thereby minimizing the risk of achieving lower levels of reliability and maintainability than anticipated. One of the more important areas of concern during a system design review involves configuration item interfaces. When two or more configuration items with known performance characteristics, are combined to produce a unique configuration, it is simple to assume that the functional performance characteristics exhibited by the combination, will transpire some logically predictable, dependent function. Because such straightforward results rarely come about when the theory is subjected to practice, increased emphasis must be accorded by management to the potential problems of functional interfaces during the system design review.

12.3.3. Preliminary Design Review

A preliminary design review will normally be carried out for each configuration item or combination of configuration items. Factors such as system complexity and the degree of difficulty experienced by the contractor in transforming the theory underlying system design into successful applied technology, will be directly reflected in the number of preliminary design reviews deemed necessary during a system's design and development.

The objectives of a preliminary design review are, (1) to establish design adequacy and determine the risk and cost associated with proposed design methods and techniques, (2) to reconcile the proposed design with the specifications of selected configuration items, and (3) to ensure that configuration items are physically and functionally compatible with one another, and with their intended operating and maintenance environments.

Preliminary design review may encompass the analysis of a wide range of configuration items up to, and including, the entire system. Similarly, the detail available to facilitate any particular review can include any, or all, of the following: schematic diagrams and drawings, simulation models, breadboard models, mock-ups, and full-scale prototypes. To ensure that the delivered system meets or exceeds the required reliability and maintainability values, all possible sources of statistical detail must be researched to the extent warranted by the existing degree of confidence in the anticipated physical and functional performance characteristics.

12.3.4. Interim Design Review

Formal interim design reviews will be conducted during a system's development. It is conventional practice for the prime contractor to be made responsible for the scheduling of all interim design reviews under the terms of a development contract (including those held to review the design of configuration items furnished by sub-contractors).

Reliability growth and maintainability improvement programs will be coordinated with the interim design review schedule, such that the reliability and maintainability goals specified in the contract will be taken into account at predetermined points in the system's development profile.

In general terms, it is the basic purpose of interim design reviews to ascertain that specified milestone events in a system's development, are being completed in accordance with an approved schedule and to specified standards. To furnish these assurances, reliability and maintainability tests and demonstrations will normally be conducted to support the interim design review process. If the results of the tests and demonstrations conform to the specified requirements, the related interim design review will be considered successful, and work can proceed on the next scheduled state of system development.

12.3.5. Critical Design Review

When a prototype of a configuration item, or any combination of configuration items, has successfully met all test and demonstration requirements, and the design is considered ready to be frozen, a critical design review will be held. Management should ensure that one or more designated customer representatives witness the qualification procedures, and ascertain in particular, that reliability and maintainability tests and demonstrations are conducted under precisely the conditions specified by the contract. Once all of the prescribed tests and demonstrations have been carried out, either approval will be given for the manufacture of additional prototypes, or full-scale production will be authorized, depending on the system complexity and the need for formal qualification trials under actual operating conditions.

12.3.6. Formal Qualification Review

A formal qualification review is the final engineering review to precede full-scale production. The term is popular in military parlance and can sometimes be considered as synonymous with critical design review when the latter leads directly to full-scale production. The factor with the greatest influence on the choice of terminology will be the complexity of the system under development and, in turn, the number of critial design reviews leading up to formal acceptance of the system.

A formal qualification review will be accompanied by a combination of test and inspection activities designed to confirm that the entire system meets the functional and physical specifications contained in the contract. Reliability and maintainability demonstration tests, together with an analysis of test results, constitute a critically important part of the formal qualification review.

12.4. DESIGN CHANGES, DEVIATIONS AND WAIVERS

Design changes, deviations and waivers have distinctly different definitions, and each is situated to a different set of circumstances. A design change is a permanent alteration to the design specification of an item prior to acceptance of the specification by the customer and, by extension, prior to the specification's use for manufacturing purposes. By definition, then, a design change takes place prior to the critical design review for a given item. The most common purpose of a design change is to improve reliability, maintainability, capability, or to reduce costs.

A design deviation is a temporary departure from a specified functional or design requirement, for a stipulated time period, or for a number of manufactured items. A design deviation is authorized on the understanding that the contractor will resume compliance with the original specification upon expiry of the stipulated conditions. The requirement for a design deviation would most often arise due to difficulties with manufacturing processes, or the unavailability of one or more specific parts for which nonpreferred substitutes are available.

The actual impact of a design deviation upon reliability, maintainability, capability and/or cost will help determine whether retrofit action is necessary to restore the items to their original status prior to the authorized design deviation. By definition a design deviation applies during the period between start of production and the delivery of the final unit.

A design waiver is formal (i.e. written) authorization to accept one or more items found during production or post-production inspection, that depart from specified requirements, but nevertheless, are suited for use, with or without some type of approved reworking. Use of the waiver procedure is again meant to overcome a transitory condition and has no effect on reliability, maintainability, capability, or costs.

12.5. ENGINEERING CHANGE

Engineering change is the term applied to any alteration applied to an item after its configuration identification has been formally established. Such items range from those under development to those which have already been delivered to the customer. The recognized method of describing and proposing an engineering change is through initiation of a document known as an engineering change proposal or sometimes called an engineering change order.

An engineering change proposal should include a clear statement of purpose, such as improvement of operating characteristics to circumvent parts shortages and to improve reliability and/or maintainability. The proposal should be accompanied by a quantitative estimate of the impact of the proposed change on reliability and maintainability. This estimate will be used by life cycle material managers as a basis for the analysis of changes to inventories, manpower, support equipment, and other logistic resource factors, all of which impact upon life cycle cost.

12.6. CONCEPTION STAGE REVIEWS

The conception stage derives its name from the fact that the system which will best meet an operational requirement, has not been identified or, as is frequently the case, does not yet exist. Thus, material managers must "conceive" the best solution to the requirement. Given a fixed budget to invest, a wide range of conceived solutions must be examined. For example, an extremely reliable system may be prohibitively expensive to produce, and one that is economical to produce may be prohibitively expensive (and time-consuming) to maintain. The system ultimately selected will typically fit somewhere within the limits signified by the two examples. One of the prime objectives of a system requirements review is to assist material managers with the selection process. Some of the concerns to be resolved in the course of this process are

1. The relative importance of various specified requirements, with the possible elimination of, or adjustment to, those which may not be achievable, or those which would entail a prohibitively large investment.
2. The potential impact on the reliability and maintainability of an increasingly complex system, resulting from a complicated, multi-functional set of requirements.
3. The viability of providing effective logistics support for the eventual system.

12.7. ACQUISITION STAGE REVIEWS

The remainder of the design reviews defined and described in Section 12.3 will all take place during the acquisition phase. As a new system evolves from the design specifications, all the way to acceptance, and then to the completely assembled system during the acquisition phase, design reviews permit material managers to monitor the progress of the system's development. Additional important functions of the design reviews are to furnish the feedback information necessary for corrective action to take place (by design deviation, waiver, change/engineering change), and to aid in the formulation and clarification of design objectives to be reached at successive reviews.

Central to the conduct of a successful design review is the application of configuration management principles. Configuration management involves the identification, control, and status accounting of all configuration items (i.e. items designated to perform an explicit operational function as part of a system and denoted as a configuration item for material management purposes). The supporting media for applied configuration management are the drawings, the data lists, the publications, and the other records required for the manufacture, the assembly, and/or the testing of a system and its components. The physical and functional "audits" carried out to ascertain that configuration items meet specified requirements are normally scheduled as a prerequisite or adjunct to the formal qualification review (see Chapter 8).

12.8. IN-SERVICE STAGE REVIEWS

Although the design review process is formally completed prior to the start of the in-service phase, there is widespread recognition of the need for material managers to periodically review a system's effectiveness under actual operational conditions. Reliability and maintainability demonstration tests normally take place under conditions that tend to be somewhat structured and controlled, when compared to the unanticipated environmental stresses imposed by in-service operation. Also, there is a moderate to high likelihood that production defects will occur in parts of a new design or, because of the unique interfacing of proven and reliable parts, innovative functional characteristics will be observed.

The aim of in-service reviews is to isolate the individual causes of frequent system failure and high mean repair times, and to develop and institute modifications to improve operational reliability and maintainability. An effective management information system is an essential requirement for the collection and interpretation of the statistical information needed for reliability and maintainability monitoring (see Chapter 11). The demonstrated reliability and maintainability levels, upon which system acceptance is based, as is objective criteria for the measurement of system effectiveness. Although the demonstrated levels are seldom achieved under operational conditions (especially complex systems subjected to severe environmental stresses), an estimate of reasonable ratios of demonstrated to achieved reliability and maintainability can be formulated by the extrapolation of management information relating to other systems of similar type and technology.

12.9. DISPOSAL STAGE REVIEWS

Whereas the reviews that take place during the first three life cycle phases are all designed to achieve or improve upon specific levels of reliability and maintainability, disposal phase reviews will normally be carried out for the purpose of establishing future objectives through an analysis of failure and repair statistics on the system to be phased out.

Extrapolation can again be employed by comparing the reliability and maintainability improvements achieved from the replacement of other similar systems, with new systems of more advanced technology, in order to establish an "order-of-magnitude" ratio of improvement. This ratio can then be used as a starting point for the development of specifications for a replacement system.

The disposal phase also affords the opportunity to review the entire logistics history of the system to be replaced. The objectives here, are to refine and improve upon the maintenance concept, thereby optimizing future logistics resource investment and application decisions.

12.10. WAIVER/DEVIATION AND ECP FORMS

Two types of forms should be considered for recording waiver and deviation requests and engineering change proposals. These two forms can assist management in making decisions. The following information should be considered for such forms.

For the waiver and deviation request form, the information may include:

- Name of originator.
- Criticality of deviation or waiver.
- Type designation of item by system or document.
- What is effected (e.g., function, product, etc.).
- Documents effected, including drawings.
- Title of waiver or deviation.
- Effect on cost.
- Effect on interfaces, spares, etc.
- Description of waiver or deviation.
- Reason for waiver or deviation.
- Approval signatures.
- Date.

For the engineering change, the form may contain some of the following information:

- System, assembly, and item number affected.
- Hardware or software effected.
- Related documents.
- Title of engineering change.
- Reason for change.
- Effect on cost.
- Effect on interfaces.
- Description of change.
- Action responsibilities.
- Distribution of form.
- Approval signatures.
- Date.

12.11. DESIGN REVIEW INPUT

For a design review to be successful, specific types of input information must be available. This information is obtained from a combination of specifications, drawings, management and test reports, examination of actual hardware, etc. An effective method of identifying and collecting the necessary input information is to employ a design review checklist.

A design review checklist consists of a comprehensive list of specific points to be resolved, and questions to be answered during the design review. For example, various components of the system must be categorized as either repairable or disposable. The design review checklist for the applicable design review would therefore include a list of all such components. The associated input information is typically obtained from a level of repair analysis conducted to determine whether it is more cost-effective to repair or discard each component at any given repair location. A level of repair analysis is but one of a wider range of activities that together, comprise a logistic support analysis. The reliability and maintainability characteristics of the system and its components, at each indenture level, form the basic input information, which ultimately determines the most cost-effective combination of logistics resources and the life cycle cost of acquiring, operating, and maintaining the system. More than any other factor, it is the long-term impact upon cost-effectiveness that makes accurate design review input information so necessary and important.

12.12. DESIGN REVIEW OUTPUT

The output of a design review consists of information to be used either for feedback purposes or to facilitate decisions to proceed with the next stage of system development. In general, it is feedback information that prompts the incorporations of design and engineering changes.

It is important to note that decisions leading to approval for a project to progress to its next developmental stage should be made only after an analysis of all cost implications has been completed. Both the near-term costs for further design and development work and, even more important, the impact upon life cycle cost, should be determined by management.

12.13. DESIGN REVIEW TEAMS

The manufacturer and the customer will each be represented by a team of engineers and specialists at a design review. To facilitate the question and answer process which must take place during each review, it is mandatory that engineers and specialists knowledgeable in reliability and maintainability, and in the interrelated disciplines, either directly participate in the review, or be available on short notice. Some of the disciplines with a potential impact upon, and/or affected by reliability and maintainability are system and detail design, human factors engineering, safety engineering, standardization, software and documentation, production, quality control and assurance, configuration management, logistics support, and life cycle costing.

Experience has shown that a design review will prove most effective if the manufacturer's and customer's design review teams are each composed of no more than ten representatives. Similarly, there are strong arguments for the selection of some representatives who, while qualified in their respective disciplines, are not intimately familiar with the system's design intricacies. These provisos on team size and individual knowledge are conducive to the free flow of information and to the rendering of sound and impartial decisions.

12.14. DESIGN REVIEW BENEFITS

A whole range of benefits may be expected to accrue from a well-planned and executed design review. Such benefits will in fact, far outweigh the investment in time and money necessary to carry out a design review, particularly in the long term. Some of the prospective benefits are

1. Assurance that the design and configuration comply with customer specifications.
2. Successful configuration management.
3. Resolution of reliability and maintainability interface problems.
4. Effective monitoring of developmental progress and status.
5. Optimization of maintenance planning and logistics support investment decisions.

6. Fulfilment of operational system effectiveness objectives.
7. Reduced life cycle cost.

12.15. EXAMPLE PROBLEMS

12.15.1. Engineering Change to Existing System

A complex radar antenna demonstrated an MTBF of 1,300 hours and an MTTR of 1.2 hours. A desired improvement in the radiation pattern could not be achieved without expensive redesign of the existing antenna—an uneconomical prospect in view of the high annual repair cost imposed by the noted MTBF and MTTR.

The supplier of a new antenna submitted an engineering change proposal to the user. The proposal contained a claim that the new antenna would exhibit an MTBF and MTTR of 25,000 hours and 0.35 hours respectively, although it had not been thoroughly tested under the environmental conditions associated with the user's operational requirements. Nevertheless, the new antenna was purchased and installed in all of the user's radar systems. At the end of the first year of operation, failure and repair statistics revealed an MTBF of 5,000 hours and an MTTR of 2.1 hours. An analysis of the experience uncovered the following contributing factors and their effects, which might have been avoided or minimized had the proposed change been more thoroughly researched:

1. Extreme and rapid changes in temperature which caused electrical leakage and premature failure of solid state devices.
2. Special training and support equipment were necessary to test and repair circuitry, which failed more frequently than anticipated.
3. Alignment difficulties were experienced when parts of a failed array had to be replaced.
4. An increase in preventive maintenance frequency was instituted in an attempt to reduce unanticipated failures.
5. Although an MTBF of 5,000 hours represented a marked improvement over the MTBF of the original antenna, spare parts and manpower forecasts were based on the supplier's claim of a 25,000-hour MTBF. Thus, parts shortages contributed to a higher than expected MTTR.
6. Forecast operating and maintenance funds were inadequate.

12.15.2. Specification Changes During System Development

A fleet of military aircraft were scheduled for delivery over a period of a few years. The aircraft were equipped with an extensive range of interactive electronics subsystems, which were designed to customer specifications, although the basic aircraft had been in service elsewhere for a number of years. During the design and the development of the electronics installations, the specifications were changed on several occasions, at the customer's request, in order to broaden the aircraft's operational capabilities. Firm disposal plans had been made for the aircraft, which were to be phased out of service. As a result of the design changes and unalterable replacement schedule, the customer was forced to begin taking delivery of the new aircraft before they could be properly maintained by his maintenance organization. Software programs for both the embedded computer systems and the automated test equipment had to be modified, while technical manuals required extensive amendments. Spare parts provisioning had to be repeated to accommodate hardware changes. When failures began to occur, technicians were unable to correctly identify parts, but were unable to obtain them. Fleet serviceability dropped rapidly as parts were robbed and rerobbed to keep a dwindling number of aircraft in operation. Meanwhile, the operational capability of those remaining in service was diminished until the amended technical manuals and modified software were finally delivered.

The example demonstrated some of the pitfalls of specifying design changes to sophisticated, software-reliant electronics subsystems in midstream, not to mention the increase in design and development costs. On the other hand, a potential military adversary will sometimes disclose the development of a new type of threat, so that his adversary will invest scarce resources to design and develop hardware and software to counter the threat.

There are, at times, no simple solutions to complex operational problems, beyond making a concerted effort to apply as much foresight as possible to the planning process. And although the example outlined above involves a military weapon system, it is easy to foresee that a similar series of design changes for commercial or industrial systems or equipment, could prove catastrophic to a company's survival prospects.

12.15.3. Environmental Effects of Mature Equipment

Modern teletype equipment was in service with a telephone company, and failure and repair statistics were available for a two year period. The MTBF was 1,800 hours and the Mean down time (MDT) for both corrective and preventive maintenance was 0.25 hours. These figures were well within the performance specifications set by a customer, who intended to install the equipment for shipboard use. Several sets of equipment were purchased, together with an inventory of spare parts based upon an analysis of available failure statistics.

Within a few months of shipboard operation, a sufficient data base of failure statistics was available to confirm an MTBF of approximately 1,000 hours, which is equivalent to a failure rate of 80% higher than expected. The situation also led to parts shortages and an increase in MDT. The user was then faced with the alternatives of either tolerating the impact upon operational effectiveness and support costs, or conducting a costly investigation to determine the cause(s) of an inordinately high failure rate.

A key factor to note in this example is the implicit acceptance of failure and repair statistics collected under significantly different environmental conditions. Moreover, demonstration tests, conducted as a condition of acceptance, are a more economical proposition for a customer, than are the far-reaching consequences associated with waiving the requirement for such tests.

12.16. SUMMARY

Design reviews and changes are essential to the attainment of reliability and maintainability goals. They are indispensable components of the total spectrum of engineering and maintenance management activities, which are designed to satisfy a customer's system effectiveness requirements for the minimum possible life cycle cost.

A proposal to incorporate a design change should be supported by reliable statistical evidence of the need for the change, and by the projected effect on reliability, maintainability, capability, and cost. Design improvements in reliability and maintainability may impose a heavy front-end investment, but the cost may be recovered several times over by way of logistics support savings. Increased capability invariably adds to system complexity and cost, and it may detract from reliability and maintainability. The need for added capabilities must therefore be weighed against the impact upon all other system effectiveness variables. Another frequently important factor, which must be considered before a system is accepted and deployed, is the intended operating and maintenance environment, and the related requirement to establish the integrity of the design through suitable qualification tests and demonstrations.

Throughout the system life cycle, reliability and maintainability reviews and analyses will demand a greater amount of time and effort from engineering and maintenance managers, than will any other combination of activities. Conversely, the time and effort devoted to reliability and maintainability reviews and analyses will be directly reflected by a system's cost-effectiveness.

12.17. EXERCISES

1. What are the aims of a design review?
2. Describe the circumstances under which the following may be required: (a) design change, (b) design deviation, (c) design waiver.
3. Outline the information required to substantiate an engineering change proposal.
4. Identify the engineering and project management disciplines to be represented by specialists on design review teams.
5. Consider the environmental factors that influence reliability and maintainability. Include logistics support elements in the definition of environmental factors.

CHAPTER 13

Life Cycle Costing and Warranties

13.1. INTRODUCTION

In recent years, increasing attention is being given to the life cycle cost of a system or a product. This may be due to the factors such as inflationary trends, tight economic conditions, and budget limitations. The life cycle cost of a system is the total sum of all the costs expected to occur from the birth of the system, to its death.

According to reference [1], the concept of life cycle cost (LCC) is not as new as it may appear to be. For example, in 1933 one of the first United States Government references to life cycle costing was published by the Comptroller General.

In 1961, the Building Research Institute organized a conference, in Washington D.C., entitled "Methods of Building Cost Analysis." The papers presented in this conference were directed toward life cycle cost. In 1965 the Logistics Management Institute published an influential report, entitled "Life Cycle Costing in Equipment Procurement," for the assistant Secretary of defense.[2] Furthermore, in 1970 the United States Department of Defense (DOD) published guidelines for LCC acquisitions. In addition, a key procurement policy directive known as the Directive 5000.1 was issued by the United States Department of Defense in 1971.[3,4] Another important document entitled "Life Cycle Costing Guide for System Acquisitions (Interim)" was published in 1973 by DOD.

An important step in the history of life cycle costing took place in 1975 when State Energy Policy and Conservation Act prompted almost every state to submit a LCC plan to qualify for the funds.[5] In the same year, a project, entitled "Life Cycle Budgeting and Costing as an Aid in Decision Making," was initiated by the Department of Health, Education, and Welfare.[6] In 1978 important legislation was passed by the state of Nebraska concerning LCC.

This legislation requires a life cycle cost analysis study for each state facility worth more than $50,000.[5]

In the last few years, some books on life cycle costing have also appeared, and short courses are also being taught.[1,6,7,8,9] A list of extensive references on the subject are given in reference [10].

Nowadays, increasing attention is also given to product warranties. In this case, manufacturers provide customers with written statements guaranteeing the integrity of their products. These statements spell out the responsibilities of manufacturers, if products turn out to be defective. According to a survey conducted by the Conference Board, of the 369 United States manufacturers surveyed only 1% did not have documented or written warranties on their products.[11] A list of references concerned with warranties are given in reference [10]. Both life cycle costing and warranties are described in the subsequent sections of this chapter.

13.2. LIFE CYCLE COST NEED, USE, AND PLANNING

This section briefly explains the life cycle cost need use, and planning.

13.2.1. Need and Use of Life Cycle Cost

Some of the needs and uses of life cycle cost are as follows:[9]

1. To choose among competing manufacturers, a manufacturer to produce a product.
2. To compare the costs of alternative approaches to meet a need.
3. To make equipment replacement decisions.
4. To stimulate orderly planning.
5. To have effective control over an ongoing program.

13.2.2. Information for Life Cycle Costing

Information on the following items should be included in the plan for an LCC analysis:

1. Assumptions associated with analysis.
2. LCC analysis schedule.
3. The estimating procedures to be utilized.
4. Financing for the LCC effort.
5. Ground rules for analysis.
6. Organization of the LCC effort.

In addition, questions on items such as follows are to be asked before initiating any life cycle costing study:[12]

1. Responsibility of cost analyst.
2. Impact of precision and accuracy in the estimates.
3. General cost fund limitations.
4. Objective of the estimate.
5. Available data and availability of data.
6. Treatment of uncertainties.
7. Constraints.

Just like in any other study, the input data are important in life cycle costing. Thus, questions should be asked on items such as follows:[12]

1. Bias in data.
2. Applicability of data to the study in question.
3. Comparability of data to other data.
4. Availability of data.
5. Obsolete data (i.e. to find out whether the data are obsolete or not).

13.3. DISCIPLINES AND ACTIVITIES ASSOCIATED WITH LIFE CYCLE COST ANALYSIS

This section briefly discusses separately, the disciplines needed for life cycle costing and the activities involved in life cycle cost analysis. Many disciplines are needed in life cycle costing.[9] Most of them are as follows:

1. Contracting and estimating.
2. Finance and accounting.
3. Reliability and maintainability.
4. Quality control and statistical analysis.
5. Engineering and management.
6. Manufacture engineering and logistics.

Similarly, in life cycle cost analysis, one is concerned with the activities such as follows:

1. Identifying the cost drivers.
2. Establishing discounted and escalated life cycle costs.
3. Estimating life cycle costs in constant dollars, and determining the cause and effect relationships.
4. Establishing constant dollar cost profiles and performing sensitivity analyses of life cycle costs.
5. Establishing an accounting breakdown structure and cost estimating relationships for every individual component in the LCC breakdown structure.

13.4. LIFE CYCLE COST MODELS

There are various life cycle cost models used to compute the LCC of engineering systems. A specific life cycle cost model may be useful to compute the LCC of a certain system, but may not be that useful for another system. The reasons for this are the nonavailability of cost data in the specified format, the type of system, etc. This section presents five selective life cycle cost models.

13.4.1. General Life Cycle Cost Model

To compute the life cycle cost of a system, the following equation is frequently used[13]

$$\text{LCC} = k_{rd} + k_p + k_{Os} + k_d \tag{13.1}$$

where

LCC is the life cycle cost
k_d is the cost of disposal
k_{rd} is the cost of research and development
k_p is the cost of production
k_{Os} is the cost of operation and support.

13.4.2. Department of Energy Life Cycle Cost Model

To calculate the life cycle cost of energy systems the United States Department of Energy uses the following model developed by the National Bureau of standards[14]

$$\text{LCC}_p = \sum_{i=1}^{6} C_i \tag{13.2}$$

where

LCC_p is the present worth of the sum of life cycle cost
C_1 is the present worth of salvage value
C_2 is the present worth of cost of energy
C_3 is the present worth of cost of investment
C_4 is the present worth of replacement cost type I
C_5 is the present worth of nonfuel, yearly recurring operation and maintenance cost
C_6 is the present worth of replacement cost type II.

Most of the above terms are self-explanatory. However, two of them are described in detail below.

Present Worth of Replacement Cost Type I (C_4)
This is given by

$$C_4 = \sum_{i=1}^{k} C_r \cdot F_{spw} \tag{13.3}$$

where

C_r is the replacement cost
F_{spw} is the single present worth factor
k is the years in the LCC evaluation
i is year.

Present Worth of Replacement Cost Type II (C_6)
This is given by

$$C_6 = \sum_{i=1}^{k} C_n \cdot F_{spw} \tag{13.4}$$

where

C_n is the nonrecurring, nonfuel operating maintenance cost.

13.4.3. Software Life Cycle Cost Model

This model is concerned with computing the software life cycle cost. Thus, the LCC is given by

$$\mathrm{LCC}_s = \sum_{i=1}^{7} k_i \tag{13.5}$$

where

LCC_s is the software life cycle cost
k_1 is the cost of test and integration
k_2 is the cost of installation
k_3 is the cost of analysis
k_4 is the cost of code and checkout
k_5 is the operating and support cost
k_6 is the cost of design
k_7 is the cost of documentation.

Further, breakdowns of the above cost components are given in reference [5].

13.4.4. Health Care Facilities Life Cycle Cost Model

The total life cycle cost of health care facilities is given by

$$LCC_h = \sum_{i=1}^{2} k_i \tag{13.6}$$

where

LCC_h is the health care facilities life cycle cost
k_1 is the capital
k_2 is the operating cost.

The capital includes costs associated with financing, direct construction, site preparation, land acquisition, demolition, and so on. Similarly, the operating cost is associated with fuel and utilities, operation and maintenance of electrical systems, maintenance of fire protection systems, equipment maintenance, operation and maintenance of heating systems, escalators, elevators, painting, and so on.

13.4.5. Appliance Life Cycle Cost Model

In this case, the life cycle cost is given by[5]

$$LCC_a(y) = \sum_{i=1}^{5} k_i(y) \tag{13.7}$$

where

y is the model year
$k_1(y)$ is the contractual cost
$k_2(y)$ is the retirement cost
$k_3(y)$ is the energy cost
$k_4(y)$ is the procurement cost
$k_5(y)$ is the service cost
LCC_a is the appliance life cycle cost.

13.5. USEFUL MODELS TO ESTIMATE COST AND RETURN ON INVESTMENT

This section presents nine selective mathematical models to estimate cost and return on investment.

13.5.1. Model I

This model is known as the Lang's model. Lang proposed this method for the application in the chemical industry by reasoning that the installed cost of a plant is several times the procurement cost of the major items of equipment.[15,16] According to Lang, the installed plant cost, PC, is given by

$$PC = \alpha \cdot BC \tag{13.8}$$

where

- BC is the base cost
- α is known as the Lang factor. This factor depends on the type of plant. For example in reference [15], the values of α are 3.6, 4.7, 3.1 for a solid-fluid-processing plant, a fluid-processing plant, and a solids-processing plant, respectively.

13.5.2. Model II

This model was developed by W. Hahn and J. Stone and is concerned with computing cost of transferring software from one computer system to another.[17] Thus, the total cost of transfer, CT, is given by

$$CT = \sum\theta \cdot \frac{(1 + f_1)r_t + r_c(1 + f_2 + f_3)}{r_c \cdot r_t} \cdot k_x + k_d + \sum k_m + \sum k_{ma} \tag{13.9}$$

where

- k_x is the man day cost
- k_d is the data transferring cost
- k_m is the machine-assisted transfer methods cost
- k_{ma} is the miscellaneous cost. For example, cost of training, management, facilities, and so on.
- r_t is the base line test production rate[17]
- r_c is the baseline conversion rate[17]
- f_i is the ith factor, for i = 1 (system integration degradation), i = 2 (documentation degradation), and i = 3 (program instability degradation)
- θ is the number of instructions to be transferred from one system to another manually.

13.5.3. Model III

This model is concerned with estimating the cost of machining/piece.[15] Thus, the cost, C, per piece is given by

$$C = (p \cdot \alpha + \beta)/p \tag{13.10}$$

where

p is the pieces produced with a single tool
β is the cost of tool (this cost includes tool grinding and changing costs)
α is the machining and work-handling costs per piece.

13.5.4. Model IV

This model is used to calculate the total cost of a machined part. Thus, the total cost is given by[15]

$$TK = \sum_{i=1}^{3} k_i \tag{13.11}$$

where

TK is the total cost of a machined part
k_1 is the material cost
k_2 is the tooling cost
k_3 is the machining cost.

The values of k_2 and k_3 are estimated from the following equations, respectively

$$k_2 = k_t\, T/L \tag{13.12}$$

where

L is the life of the tool in minutes
T is the machining time in minutes
k_t is the cutting tool cost

and

$$k_3 = \frac{(T_1 + T_2)}{60} \cdot \{(A + B)/100\} \tag{13.13}$$

$$A \equiv k_m(1 + r_m) \tag{13.14}$$

$$B \equiv r_l(1 + r_O) \tag{13.15}$$

where

T_1 is the machining time
T_2 is the idle time or the nonproduction time
k_m is the machine cost in dollars/hour
r_m is the machine overhead rate in percentages
r_O is the operator overhead rate in percentages
r_l is the operator labor rates in dollars/hour.

13.5.5. Model V

This model is also used to estimate the cost. The model expresses the cost-capacity relationship. The model is used in a situation to estimate the cost of a new plant, where the capacity and cost of a similar old plant, and the capacity of the new plant under consideration are known. Thus, the cost, C_n, of the new plant can be estimated from the following equation.

$$C_n = C_O \cdot \left(\frac{P_n}{P_O}\right)^m \tag{13.16}$$

where

- m is known as the cost-capacity factor. Its value varies from about 0.4 to 0.8. However, in many cases the frequently used value for m is 0.6[18]
- P_n is the capacity of the new plant
- P_O is the capacity of the old plant
- C_O is the cost of the old similar plant.

EXAMPLE 13.1

The cost of a steel plant with a capacity of 2 million ton/year was 200 million dollars. What would be the estimated cost of a similar 3 million ton/year plant. (Assume the value of the cost-capacity factor is equal to 0.8.)

In this example, the data are specified for the following items

$$m = 0.8, \quad C_O = 200 \text{ million dollars}, \quad P_n = 3 \text{ million ton/year}, \quad \text{and} \quad P_O = 2 \text{ million ton/year}.$$

By substituting the above data into equation 13.16, we get

$$C_n = 200\left(\frac{3}{2}\right)^{0.8} = 276.632 \text{ million dollars.}$$

Thus, the estimated cost of the new plant will be approximately 277 million dollars.

13.5.6. Model VI

This model is used to estimate the integrated circuit manufacturing cost/chip. The integrated circuit manufacturing cost is given by[19]

$$C_m = C_p + C_t + C_{pc} \tag{13.17}$$

where

C_m is the integrated circuit manufacturing cost/chip
C_{pc} is the cost of packaging
C_t is the cost of the test
C_p is the cost of processing.

The cost of the test is given by

$$C_t = \frac{C_f}{T_y} \tag{13.18}$$

where

T_y is the test yield
C_f is the cost of testing finished a integrated circuit.

In addition, the costs of packaging and processing are given by, respectively

$$C_{pc} = \frac{C_{ic} + C_{cn}}{P_y \cdot T_y} \tag{13.19}$$

where

P_y is the packaging yield
C_{cn} is the integration circuit connections cost
C_{ic} is the integrated circuit package cost

and

$$C_p = \frac{C_w}{C_{cw} \cdot w_y \cdot P_y \cdot T_y} \tag{13.20}$$

where

C_w is the processing wafer cost
w_y is the wafer yield
C_{cw} is the chips/wafer.

13.5.7. Model VII

This model is concerned with estimating the satellite system launch cost.[20] Some of the assumptions associated with this model are

1. Costs (in 1974 dollars).
2. Satellite is launched by utilizing the Hohman transfer method.

Thus, the launch cost is estimated from the following equation

$$C_{lu} = 0.026\, w^{2/3} \left(\frac{LT}{1000} + \alpha + 1\right) \tag{13.21}$$

where

C_{lu} is the total single-satellite system launch cost in million dollars
α is the payload sophistication factor
LT is the orbit altitude (statute miles).
w is the satellite weight (pounds).

13.5.8. Model VIII

This model is used to compute the software development cost.[21] The following equation is used to estimate such cost

$$C_{sd} = \theta.R + \theta\, R f \qquad (13.22)$$

where

C_{sd} is the software development cost
θ is the manpower needed (man-months)
f is the ratio of secondary cost to primary cost. The secondary cost includes documentation, production, computer time, and so on.
R is the mean labor rate (dollars/man-months).

13.5.9. Model IX

This model is concerned with estimating the computer hardware, maintenance cost.[22] Thus, the maintenance cost/month is given by

$$M_c = M_{cc} + I_c + M_{pc} \qquad (13.23)$$

where

M_c is the maintenance cost/month
M_{cc} is the cost of corrective maintenance.
I_c is the cost of inventory
M_{pc} is the cost of preventive maintenance.

The cost of preventive maintenance is calculated from the following equation

$$M_{pc} = r\, h_O(t_1 + t_2)/t_3 \qquad (13.24)$$

where

t_1 is the traveling time of the customer engineer for preventive maintenance
t_2 is the scheduled customer engineer's preventive maintenance time
t_3 is the scheduled preventive maintenance time interval

h_O is the monthly operating hours of equipment
r is the hourly rate of the customer engineer.

Similarly the inventory cost and corrective maintenance cost are given by

$$I_c = i_r \cdot v \tag{13.25}$$

where

v is the value of the maintenance spare part inventory
i_r is the inventory cost rate/month.

and

$$M_{cc} = r \cdot h_O(t_4 + t_5)/t_6 \tag{13.26}$$

where

t_4 is the mean time to repair
t_5 is the traveling time of the customer engineer for corrective maintenance
t_6 is the mean time between failures.

Finally, the customer engineer's rate/hour is given by

$$r = \frac{\beta(1 + R_O)}{b} + C_p \tag{13.27}$$

where

C_p is the part cost/hour
β is the pay/hour (of the customer engineer)
R_O is the overhead rate
b is the fraction of customer engineer's time being spent on maintenance.

13.5.10. Model X

This model is used to estimate the return on investment.[22] Thus, the return on investment is given by

$$R_{Oi} = \frac{m_f \cdot C_f}{C_r(1 + \alpha_1)(1 + d)} \tag{13.28}$$

where

R_{Oi} is the return on investment
α_1 is the general and administrative overhead
d is the profit or fee

m_f is the annual number of failures avoided due to the reliability program

C_f is the repair cost of a failure

C_r is the reliability program cost. This cost includes the labor overhead cost.

13.6. STEPS FOR A CONTRACTOR TO FULFILL LCC REQUIREMENTS IN CONTRACTS

The past number of years have witnessed the emphasis of LCC in government and other procurement contracts. The LCC guarantees for the field reliability of the equipment, its logistics support cost, or contractor maintenance, in the form of reliability improvement warranties.[24] The LCC requirements have created a new commitment for contractors or manufacturers. The contractor or the manufacturer can satisfy the LCC commitments of the contract only through careful planning and proper measures. This section presents a seven step approach for contractors to meet LCC requirements committed in a contract.[24] The procedure is shown in Figure 13.1.

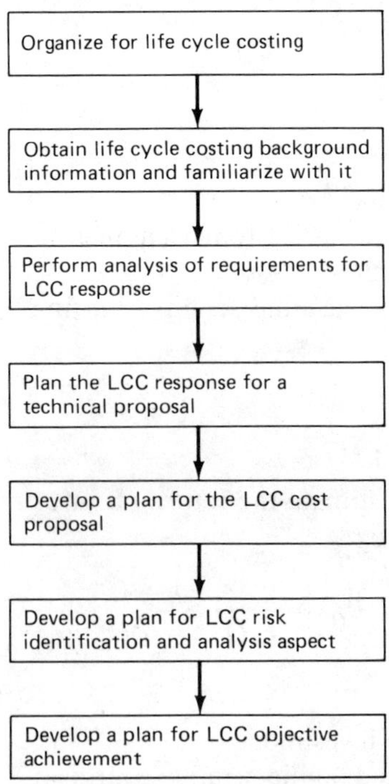

Figure 13.1. A seven step approach for contractors to meet LCC requirements.

(The steps shown in Figure 13.1 are not described in detail because they are considered to be self-explanatory.)

13.7. EVALUATION OF LIFE CYCLE COSTING PROGRAM

In order to have an effective life cycle costing program, the management should evaluate its program periodically. The evaluation questions should pertain to the following areas:

1. Qualification of the life cycle costing representative.
2. Broadness of the data base for cost estimating.
3. Availability of estimates to support design groups.
4. Awareness of LCC goals by the planning and design groups.
5. Identification of cost drivers.
6. Construction of a cost model.
7. Periodic evaluation of subcontractors' cost performance.
8. Validation of cost estimates by an independent appraisal.
9. Soundness of cost estimating techniques.
10. Validity of assumptions associated with cost estimates.
11. Compatibility of safety, reliability, and maintainability with the LCC needs.
12. Trade-off studies.
13. Consideration of inflation and discounting factors.
14. Incorporation of life cycle costing requirements in design subcontracts.

13.8. ADVANTAGES AND DISADVANTAGES OF LIFE CYCLE COSTING AND USEFUL GUIDELINES FOR LCC

This section presents the advantages and disadvantages of life cycle costing and useful guidelines for LCC. Some of the advantages and disadvantages are as follows:

Advantages

1. Equipment acquisition based on minimizing life cycle costing is useful in reducing the total cost.
2. LCC is useful to control programs.
3. LCC is useful in making decisions associated with equipment replacement, budgeting, and planning.
4. LCC is useful to compare the cost of competing projects.
5. An LCC plan is useful in providing appropriate guidance for the tasks of analysis.
6. LCC becomes useful when making a selection among the competing contractors.

Disadvantages

1. LCC is time consuming.
2. LCC is an expensive concept.
3. Collecting data for LCC analysis is a trying task.
4. The accuracy of data is doubtful.

A list of guidelines for life cycle costing is as follows:[9,25]

1. The life cycle costing unit should be attached to the financial or logistics organizations.
2. The life cycle costing concept is gaining recognition.
3. The life cycle costing must start as early as possible in the program.
4. Performance requirements should be documented when acquisition is based on the life cycle cost.
5. Both the management of the customer and the manufacturer should be capable of using LCC effectively.
6. Necessary time should be allowed to the life cycle costing group to perform analysis.
7. Necessary cost data should be available to the LCC group.
8. Return from the life cycle cost application to nonrepairable products is quite significant.
9. Maximum potential of life cycle costing lies in the early acquisition of large systems.
10. As the material level descends from large systems through subsystems and repair components, the future savings from the application of the LCC approach diminishes.

13.9. USEFUL INTEREST FORMULAS FOR LIFE CYCLE COSTING

This section presents six interest formulas, which are useful in life cycle costing. Derivations of these formulas may be found in books on engineering economy.

FORMULA I

This is known as the single-payment, compound-amount formula. In this case, the future sum of money, s_f, is given by

$$s_f = P(1 + i)^k \tag{13.29}$$

where

P is the original principal sum, at the present time
i is the annual compound interest rate

k is the number of (interest) conversions periods (usually 12 month periods).

FORMULA II

This is known as the single-payment value formula. The present value of s_f from equation 13.29 is given by

$$P = s_f(1 + i)^{-k} \tag{13.30}$$

FORMULA III

This is known as the equal-payment series, compound-amount formula. In this case, the future sum of money is given by

$$s_f = M \frac{(1 + i)^k - 1}{i} \tag{13.31}$$

where

M is the payment deposited at the end of each period.

An example of this situation is $2,000 deposited at the end of each period (i.e. year) for, say, 10 years at the compound interest rate of 10%.

FORMULA IV

This is known as the equal-payment series, sinking-fund formula. Thus from equation 13.31 we get

$$M = s_f \left[\frac{i}{(1 + i)^k - 1}\right] \tag{13.32}$$

FORMULA V

This is known as the equal-payment series, present-value formula. The formula is concerned with calculating the present value of the equal payment made at the end of each coming k period (i.e. years). Thus, in this case the present worth, P_w, is given by

$$P_w = M \left[\frac{(1 + i)^k - 1}{i}\right] (1 + i)^{-k} \tag{13.33}$$

13.10. WARRANTIES

In today's environment, the product manufacturers usually provide some kind of written document to their buyers, which guarantees the integrity of their manufactured items. Furthermore, these documents describe the responsibilities of the manufacturers, if their products happened to be defective. Nowadays, the use of warranties is so wide spread that the survey conducted by the Conference Board indicates that out of the 369 United States manufacturers, only 1% had no documented or written warranties on their manufactured products.[11] In addition, these manufacturers cited the following reasons for providing warranties:

1. Competition.
2. Defining liabilities and warranties.
3. Providing support for their produced goods.

The following are some of the reasons for having a warranty:[26]

1. It is useful to have for prompt acceptance of a manufactured product.
2. It is useful because it helps to expedite payment for the sold item from the buyer.
3. It is useful to encourage the producer to manufacture reliable and good quality products.
4. It is useful to increase the marketability of the manufactured items.
5. It is useful to place the responsibility on product manufacturers for rectifying faults in their products.
6. It helps to assure customers that their procured products will at least fulfill their specified contractual requirements.
7. It compels the producers to provide better product support documentation and services.

13.10.1. Warranty Obligations to Electrical and Mechanical Product Manufacturers

There are many options opened to electrical and mechanical product manufacturers to fulfill their warranty obligations.[11] Some of them are as follows:

1. Replace the defective parts with the new ones.
2. Replace the entire defective product.
3. Pay for the service labor to rectify faulty parts.
4. Pay customers so they can procure replacements for the faulty parts.

13.10.2. Airline Avionics Equipment Warranty Features

This section presents an example of warranty feature in the airline business. The airline avionics equipment warranty has various features.[27] Some of these are shown in Figure 13.2.

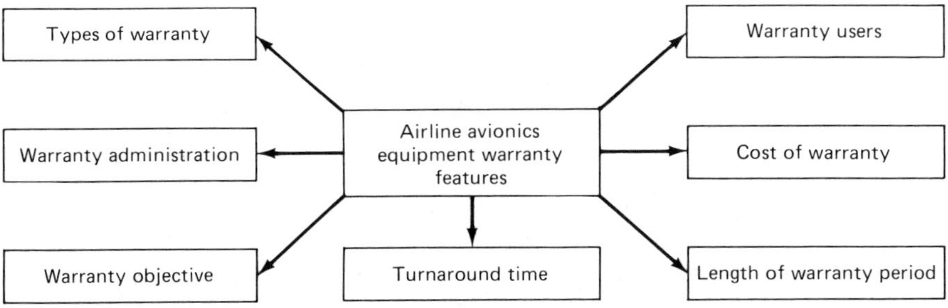

Figure 13.2. Airline avionics equipment warranty features.

13.10.3. Life Cycle Cost Model for Avionics Equipment Procurement

This model takes into consideration the equipment warranty. Thus, the life cycle cost is given by

$$C_{Lt} = k_m + k_{rs}t + \left(\sum_{i=1}^{2} k_i\right)(1 + f)f_r(t) + m \cdot P \cdot f_a \qquad (13.34)$$

where

- k_m is the user direct maintenance costs with a warranty procurement
- k_{rs} is the recurring support costs/month with a warranty procurement
- t is the calendar time period in months
- C_{Lt} is the life cycle cost over time period t for a warranty procurement
- k_1 is the cost to manufacturer for modification, discounted and amortized with a warranty procurement
- k_2 is the direct warranty repair costs to manufacturer
- f is the fee of the manufacturer
- $f_r(t)$ is the factor of risk, which the manufacturer uses to cost for a warranty time period, t, and is given by $(1 + j)^{t/12}$ for $0 < j < 1$.
- m is the number of items procured with a warranty procurement
- P is the acquisition price of each item
- f_a is the factor of amortization for time period t and is given by t/t_e. Where t_e is the expected life of equipment.

13.10.4. Standard Warranty Coverage for an Aircraft

This section presents an example of warranty coverage for a DC-10 aircraft.[28] Most of the items included in the warranty are shown in Figure 13.3.

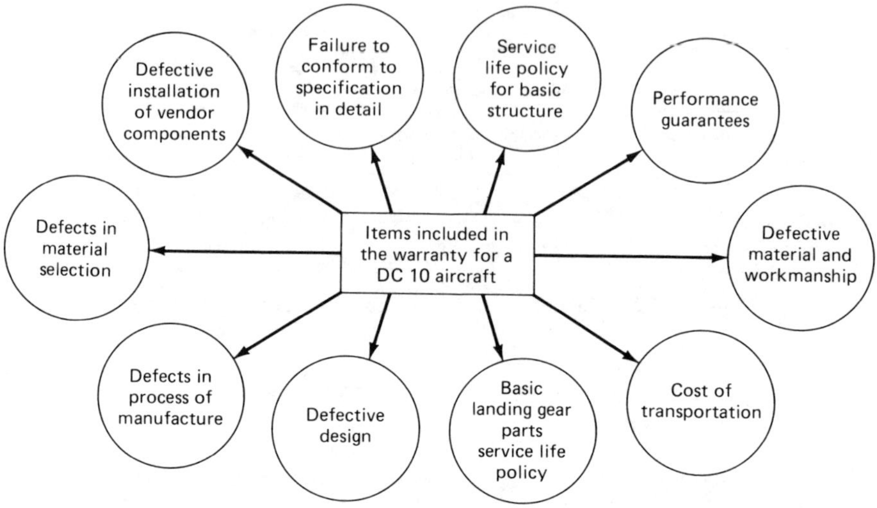

Figure 13.3. Items included in the aircraft warranty.

13.10.5. Reliability Improvement Warranty (RIW)

This is a means of encouraging manufacturers to produce systems with low cost of repairs and low failure rates. In addition, all the RIW contracts contain incentives for manufacturers.

Taking into consideration various contingencies, there are several different ways and means for structuring reliability improvement warranty terms and conditions. One frequently used provision in the RIW contract is the guarantee from the manufacturer that the equipment mean time between failures (MTBF) will prove to be accurate in the field environment. If the specified MTBF is not satisfied, the manufacturer must provide relief.

Some of the manufacturer reliability improvement warranty risk factors are as follows:[29]

1. Uncertainty in the reliability prediction.
2. Precise usage rate of item is not known.
3. Data is not supplied to the manufacturer according to the requirements.
4. Notification of intention to use warranty is late.
5. Failures outside the control of manufacturer.
6. For long-term agreements, uncertainty associated with the rate of inflation.
7. Application of RIW is concerned with advanced technology.
8. For engineering change proposal approval, procedures are time consuming.

For each of the above item, the risk reduction procedures are outlined in reference.[29]

13.11 SUMMARY

This chapter briefly presents the topics of life cycle costing and warranties. The chapter begins by discussing the life cycle cost need, and use and planning are described. In addition, the disciplines and activities associated with life cycle cost analysis are discussed. The topic of life cycle cost models is explored. Five life cycle cost models are presented, which are associated with health care facilities, appliances, software, and so on. In addition, the chapter contains nine mathematical models to estimate cost and return on investment. Eight of these models deal with cost estimation in cases such as transferring software from one computer system to another, machined part, manufacture of integrated circuits, satellite system launch, computer hardware maintenance, software development, etc.

A seven step approach for contractors to meet life cycle cost requirements is outlined. To evaluate the life cycle cost program, a check list of questions is presented. Useful guidelines for life cycle costing are given, along with advantages and disadvantages of life cycle costing. Five interest formulas useful for life cycle cost analysis are presented.

Warranties are another major topic covered in the chapter. Reasons for having warranties are outlined, along with warranty obligations to electrical and mechanical product manufacturers. Airline avionics equipment warranty features are outlined. A life cycle cost model for avionics equipment with a warranty procurement is presented. Standard warranty coverage for a DC 10 aircraft is briefly discussed. The reliability improvement warranty is the last topic discussed in the chapter.

13.12. EXERCISES

1. Discuss the following three terms:
   i. Life cycle cost.
   ii. Reliability improvement warranty.
   iii. Warranty.
2. Discuss the historical developments associated with the subject of life cycle costing.
3. Prepare a list of questions to be asked before initiating a life cycle cost study.
4. What are the disciplines that provide input in a life cycle costing study?
5. What are the benefits and drawbacks of life cycle costing?
6. Describe the term return on investment.
7. Explain the following two terms:
   i. Equal-payment series compound-amount.
   ii. Equal-payment series sinking-fund.

8. What are the reasons for having a warranty for a product?
9. What is the standard warranty coverage for a commercial aircraft?
10. Outline at least seven contractor reliability improvement warranty risk factors.
11. A 2500-MW thermal power generating station costs $500 million to build. What would be the estimated cost of a similar 3000-MW power plant, if the value of the cost-capacity factor is 0.7?
12. A person wishes to put a $10,000 down payment on a house to be purchased after a five-year period. He desires to deposit a series of uniform, year-end payments over the five-year specified period. If the estimated compound interest rate is 10%/annum, calculate the amount that should be deposited each year.

13.13. REFERENCES

1. Dell'Isola, A. J., Kirk, S. J. *Life Cycle Costing for Design Professionals.* McGraw-Hill Book Company, New York, 1981, pp. 4–5.
2. Logistics Management Institute. *Life Cycle Costing in Equipment Procurement, LMI Tasks 4C-5 Report.* Washington, D.C., April, 1965.
3. Department of Defense. *Acquisition of Major Defense Systems, DOD Directive* No. 5000.1. Washington, D.C., July, 1971.
4. Earles, D. R. LCC-Commercial Application Ten Years of Life Cycle Costing. Annual Reliability and Maintainability Symposium. 1975 pp. 74–85. Available from the IEEE.
5. Earles, M. E. *Factors, Formulas, and Structures for Life Cycle Costing.* Mary Eddins Earles, 89 Lee Drive, Concord, Mass. 01742, 1978.
6. Navamore, Bain, Brady, and Johnson. "Life Cycle Budgeting and Costing as an Aid in Decision Making, HRP-0011883." Report No. 2-Processes and Concepts, Seattle, Washington, December 1975.
7. Brown, R. J., Yanuck, R. R. *Life Cycle Costing: A Practical Tool Guide for Energy Managers.* Fairmont Press, Inc., Atlanta, Georgia, 1980.
8. Blanchard, B. S. *Design and Manage to Life Cycle Cost.* M/A Press, Portland, Oregon, 1978.
9. Seldon, R. M. *Life Cycle Costing: A Better Method of Government Procurement.* Westview Press, Boulder, Colorado, 1979.
10. Dhillon, B. S. "Life Cycle Cost: A Survey." *Microelectronics and Reliability: An International Journal.* vol. 20, 1980, pp. 737–742.
11. McGuire, E. Patrick "Industrial Product Warranties: Policies and Practices." A Research Report from the Conference Board, 1980. Available from the Conference Board, Inc., 845 Third Avenue, New York, N.Y. 10022.
12. Reiche, H. "Life Cycle Cost." In *Reliability and Maintainability of Electronic Systems,* edited by J. E. Arsenault and J. A. Roberts. Computer Science Press, Potomac, Maryland 20854, 1980.
13. Dhillon, B. S. *Reliability Engineering in System Design and Operations.* Van Nostrand Reinhold Company, New York, 1983, 210–238.
14. United States Department of Commerce. *Life Cycle Cost Manual for the Federal Energy Management Program.* National Bureau of Standards Handbook 135, Washington, D.C., 1980.

15. Dieter, G. E. *Engineering Design: A Materials and Processing Approach.* McGraw-Hill Book Company, New York, 1983.
16. Woods, D. R. *Financial Decision Making in the Process Industry.* Prentice-Hall, Inc., Englewood Cliffs, N.J., 1975.
17. Hahn, W., Stone, J. "Software Transfer Cost Estimation Technique." Report No. M70-43, MITRE Corporation, July 1970.
18. Chilton, C. H. "Six-Tenths Factor Applies to Compute Plant Costs." *Chemical Engineering,* Vol. 57, 1950, pp. 112–114.
19. Phister, M. *Data Processing Technology and Economics.* Santa Monica Publishing Company, Santa Monica, CA, 1979.
20. Hadfield, B. M. "Satellite-Systems Cost Estimation." IEEE Transactions on Communications, Vol. Com-22, No. 10, October 1974.
21. Doty, D. L., Nelson, P. J., Stewart, K. R. *Software Cost Estimation Study,* RADC-TR-77-200. Doty and Associates, Inc., L1977. Available from the National Technical Information Service, Springfield, VA.
22. Phister, M. "Analyzing Computer Technology Costs-Part II Maintenance." *Computer Design,* October 1978.
23. Coppola, A. "Reliability as a Capital Investment." Proc. of the Annual Reliability and Maintainability Symposium, 1972, pp. 349–357.
24. Schmidt, B. A. "Preparation for LCC Proposal and Contracts." Proc. of the Annual Reliability and Maintainability Symposium, 1979, pp. 62–66.
25. Ryan, W. J. "Procurement Views of Life Cycle Costing." Proc. of Annual Symposium on Reliability and Maintainability 1979, pp. 164–168.
26. Flottman, W. W., Worstell, M. R. "Mutual Development, Application and Control of Suppliers Warranties." American Airlines and Litton Systems-Aero Products Division, Proc. of the Annual Reliability and Maintainability Symposium, 1977, pp. 213–221.
27. Balaban, H., Retterer, B. "The Use of Warranties for Defense Avionics Procurement." Proc. of Annual Reliability and Maintainability Symposium, 1974, pp. 363–368.
28. Bayer, H., Speir, R. N. "Long Term Commercial Warranty." Proc. of the Annual Reliability and Maintainability Symposium, 1978, pp. 50–54.
29. Balaban, H. S., Meth, M. A. "Contractor Risk Associated with Reliability Improvement Warranty." Proc. of the Annual Reliability and Maintainability Symposium, 1978, pp. 123–129.

Index